ISLANDS, ISLANDERS, AND THE BIBLE

Society of Biblical Literature

Semeia Studies

Gerald O. West, Editor

Editorial Board
Pablo Andiñach
Fiona Black
Denise K. Buell
Gay L. Byron
Steed Vernyl Davidson
Masiiwa Ragies Gunda
Monica Jyotsna Melanchthon
Yak-Hwee Tan

Number 77

Islands, Islanders, and the Bible

RumInations

Edited by
Jione Havea, Margaret Aymer, and Steed Vernyl Davidson

SBL Press
Atlanta

Copyright © 2015 by SBL Press

All rights reserved. No part of this work may be reproduced or transmitted in any form or by any means, electronic or mechanical, including photocopying and recording, or by means of any information storage or retrieval system, except as may be expressly permitted by the 1976 Copyright Act or in writing from the publisher. Requests for permission should be addressed in writing to the Rights and Permissions Office, SBL Press, 825 Houston Mill Road, Atlanta, GA 30329 USA.

Library of Congress Cataloging-in-Publication Data

Islands, islanders, and Bible : ruminations / edited by Jione Havea, Margaret Aymer, Steed Vernyl Davidson.
    p. cm. — (Society of Biblical Literature. Semeia studies ; Number 77)
    Includes bibliographical references and index.
    ISBN 978-1-58983-946-5 (paper binding : alk. paper) — ISBN 978-1-58983-948-9 (electronic format) — ISBN 978-1-58983-947-2 (hardcover binding : alk. paper)
    1. Bible—Criticism, interpretation, etc. 2. Islands—Miscellanea. I. Havea, Jione, 1965– editor.
    BS511.3.I85 2015
    220.60914′2—dc23   2014036279

Printed on acid-free, recycled paper conforming to
ANSI/NISO Z39.48-1992 (R1997) and ISO 9706:1994
standards for paper permanence.

Contents

Preface vii
Abbreviations ix

RumInations
    Steed Vernyl Davidson, Margaret Aymer, Jione Havea 1

First Waves

Islandedness, Paul, and John of Patmos
    Margaret Aymer..25

Building on Sand: Shifting Readings of Genesis 38 and Daniel 8
    Steed Vernyl Davidson 37

Island-Marking Texts: Engaging the Bible in Oceania
    Nāsili Vaka'uta 57

Celebrating Hybridity in Island Bibles: Jesus, the Tamaalepō
(Child of the Dark) in Mataio 1:18–26
    Mosese Ma'ilo 65

Creolizing Hermeneutics: A Caribbean Invitation
    Althea Spencer Miller 77

Gaelic Psalmody and a Theology of Place in the Western Isles
of Scotland
    Grant Macaskill 97

Islands in the Sun: Overtures to a Caribbean Creation Theology
    J. Richard Middleton..115

The Island of Tyre: The Exploitation of Peasants in the Regions of
Tyre and Galilee
   Hisako Kinukawa 135

Sea-ing Ruth with Joseph's Mistress
   Jione Havea... 147

Second Waves

Sand, Surf, and Scriptures
   Roland Boer 165

Islandedness, Translation, and Creolization
   Aliou C. Niang... 177

The Wrong Kind of Island? Notes from a "Scept'red Isle"
   Andrew Mein.. 185

Third Waves

Third Wave Reading
   Elaine M. Wainwright... 201

Thinking on Islands
   Daniel Smith-Christopher 207

Writing from Another "Room-in-ating" Place
   Randall C. Bailey... 217

Contributors...227

Index of Ancient Texts..233
Index of Modern Authors...237

Preface

This anthology rides on the waves of contextual, cultural, and postcolonial criticisms, containing readings of biblical texts by islanders who are rooted in Asia, America, Caribbean, Europe, and Oceania. It takes into account the fluidity and sandiness of island spaces, the complex richness of islandness, and the sways and grooves of islandhood. The contributors write from/upon different routes, and the aim of this anthology is to guide the flow of island hermeneutics and island studies into the currents of biblical criticisms.

Most of the chapters were delivered at sessions of the Society of Biblical Literature group Islands, Islanders, and Bible (since 2009), renamed in 2012 as Islands, Islanders, and Scriptures. The chapters come together in this anthology to give a taste of how islanders might ripple the sea of biblical interpretation. In island terms, there are three clusters of waves in this anthology:

- The first cluster contains ten chapters, each flowing in/to different currents, depths, and shores.
- The second cluster offers three engagements with a selection of the chapters, as if to break up the first cluster into three waves.
- The third cluster presents three more engagements, this time breaking up the first cluster into three other waves.

This islandish collection is therefore a conversation *in formation* (or in de- and reformation, if you prefer that line of thinking), noting that different clustering produces different meaning structures, and *in transition*.

In transition, this anthology is unfinished. Lacking is a foreword, which was asked from David Jobling, stern supporter of this kind of work who as General Editor of Semeia Studies asked for a volume on Islander criticism (as the Society of Biblical Literature tags the kind of readings offered herein). Around twenty years later, Jobling is in transition, recovering from a stroke, and this collection is looking for cover (in the eyes of readers).

On the cover is Filipino artist Emmauel Garibay's *Bagong Mundo* (*New World Disorder*, 2011). Garibay offered this reading, which ends with an invitation, by email (December 6, 2013):

> The work is a depiction of Philippine colonial history. The face looking up to the sky is the idyllic precolonial era about to be altered by the intrusion of a Spanish galleon in the sky. A subtle image of a woman holding a banana and a man holding an apple form the lips and the eyeball of the face. The man and woman are *malakas* and *maganda*, the Filipino equivalent of Adam and Eve in the creation legend.
>
> In the foreground is the resistance movements that followed. The man is Andres Bonifacio the revolutionary leader. But the roots of colonialism have been deeply embedded in our consciousness (apple in the eye) resulting in a culture of subservience (bald lady with a cross) and passive acceptance of one's sufferings (face on the lower left corner). In spite of having been politically emancipated, colonialism persists culturally and ideologically (white man's face with mustache). Thus the land is perpetually a banana republic.
>
> Feel free to add your interpretation to the image.

Abbreviations

Ant.	Josephus, *Jewish Antiquities*
BA	*Biblical Archeologist*
BARIS	BAR (Biblical Archaeological Reports) International Series
CBQ	*Catholic Biblical Quarterly*
FCB	*A Feminist Companion to the Bible*
Geogr.	Strabo, *Geographica*
HTR	*Harvard Theological Review*
Int	*Interpretation*
JBL	*Journal of Biblical Literature*
JCTR	*Journal of Christian Theological Research*
KJV	King James Version
List	*Listening: A Journal of Religion and Culture*
Nat. Hist.	Pliny the Elder, *Naturalis historia*
NIDB	*New Interpreter's Dictionary of the Bible*. Edited by Katharine Doob Sakenfeld. 5 vols. Nashville: Abingdon, 2006–2009.
NIV	New International Version
NJPS	*Tanakh: The Holy Scriptures: The New JPS Translation according to the Traditional Hebrew Text*
NovTSup	Supplements to Novum Testamentum
NRSV	New Revised Standard Version
NTBIFAN	*Notes Africaines: Bulletin d'information et de correspondance de l'Institut Français d'Afrique Noire*
OBT	Overtures to Biblical Theology
SBT	Studies in Biblical Theology
SemeiaSt	Semeia Studies
TNIV	Today's New International Version
WBC	Word Biblical Commentary
WW	*Word & World*

RumInations

Steed Vernyl Davidson, Margaret Aymer, Jione Havea

We invite readers to wade into this collection of essays and to return to the Bible when the time comes with a two-part proposition: biblical texts are like islands, and readers are like islanders. At the underside of our invitation is a double affirmation: islands are like biblical texts, and islanders are (like) readers. Our invitation and double(-crossing) affirmation problematize the assumption that "no text is an island," which is a strong gust in the sails of intertextual (see, e.g., Fewell 1992) and contrapuntal (see, e.g., Sugirtharajah 2003) readers. We also challenge the assumption that "natives [islanders] can't read,"[1] which continues to blindfold colonial and missionary agents. Many nonislanders, and (truth be told) a few islanders, think that islands and islanders are naïve, simplistic, and disconnected. They deserve some islanding and sanding (see Davidson in this volume)!

The contributors to this anthology write from the surfs and turfs of islands in Asia, America, Caribbean, Europe, and Oceania. There are more island shores and island cultures out there whose ways, voices, lives, and faces are not channeled into and through this anthology. Our aim is not to be representative but to invite a conversation on how being islanders, and the various ruminations of islandedness, condition the way we read biblical texts. Toward this aim, the essays are organized, in island terms, as three clusters of waves. The first cluster contains ten chapters in which islander scholars address different aspects of island context, islander identity, and islandic peoplehood. Hence the three parts of this opening chapter, drawing attention to, and ruminating around, island space, islandness, and islandhood.

---

1. A comment made in jest, but deeply insulting, to Jione Havea after delivering presentations at two different occasions.

The second cluster offers three engagements with a selection of the ten chapters, as if to break the first cluster into three more waves: (1) Roland Boer engages with the chapters by Steed Vernyl Davidson, Nāsili Vaka'uta, and J. Richard Middleton; (2) Aliou C. Niang engages with the chapters by Margaret Aymer, Mosese Ma'ilo, and Althea Spencer Miller; and (3) Andrew Mein winds up this cluster by engaging with the chapters by Grant Macaskill, Hisako Kinukawa, and Jione Havea.

The third cluster presents three more engagements, this time breaking up the first cluster into three other waves: (1) Elaine M. Wainwright engages with the chapters by Aymer, Ma'ilo, and Kinukawa; (2) Daniel Smith-Christopher engages with the chapters by Davidson, Miller, and Macaskill; and (3) Randall C. Bailey engages with the chapters by Davidson, Vaka'uta, and Havea.

In the second and third clusters, the "first wave" is landed then rippled differently. This islandish collection is therefore a conversation *in formation*, noting that different clustering produces different meaning structures, and *in transition*, for this collection aims to ripple rather than to establish and settle.

RUMInations

The subtitle for this anthology performs a play upon a single word to evoke several other words and concepts required for thinking from the perspective of island space. In an ironic twist, we have designated "rumi" to serve as the placeholder for consideration of island space. From our own lived experiences we have only known islands as small spaces, hardly the roomy environments that typify continental spaces. Precisely in the rupture between our lived knowledge and our choice of words lies the opportunity for our theorizing. Imagining island spaces as small and isolated ignores the evidence of roomy islands like Iceland or Madagascar, Papua or Solomon, Aotearoa or Australia. This representation falls prey to the cultural production of the desert isle sufficient for a single person to engage her or his existential challenges and adequate in size to serve as the microcosm for anthropological research. Our theoretical task requires more than just consideration of representation of spaces. This task compels an extraterritoriality that embraces the sea as integral to island spatiality. This shift forecloses the notion of island space as restricted. The theory of island space that we wish to utilize as readers of sacred texts grounds us in the island as the space of thought. In theory, read both as aspirational and practical academic engagement, we see this work as more than simply reacting to, rejecting, or recasting biblical interpretations that misunderstand or mischaracterize island space. This work serves as an entry point to thinking biblically through the island.

We remain aware that we engage this project as participants and purveyors of Western academic discourses that were at the same time being formed by island space. The challenge here lies not in whether living in island space qualifies someone to participate in this conversation. Rather, the greater disqualifier comes from our intellectual formation in Western academies steeped in their constructions of island spaces. While much of early Western cultural associations were formed from an insular perspective, that is to say, from an island perspective, with greater continental expansions, islands slowly became part of the periphery of dominant Western culture.[2] When continental space dominates the intellectual and cultural landscape, islands begin to be represented as remote, unoccupied, isolated, and importantly small. This representation of island space occurs mostly in the castaway genre seen in works such as Robert Louis Stevenson's *Treasure Island* and Daniel Defoe's *Robinson Crusoe*. Elizabeth DeLoughrey (2007, 12) notes the dominance of what she calls the Robinsonade genre by citing the publication of "500 desert-island stories" between 1788 and 1910 and the multiple printing of *Robinson Crusoe* in its first year of publication.

The representation of island space as small, isolated, deserted, and, as David Lowenthal (2007, 206) notes, despite their shape perfect circles persists in various forms. Television and movie depictions like *Castaway*, *Survivor*, or *Lost* continue to reinforce the notion of the island as uninhabitable, remote, and small. Even though this representation begins as a construct of the Western imaginary,[3] it gains widespread acceptance among island residents who deploy the representation strategically for tourism purposes. As Lowenthal indicates, some of the most densely populated areas on earth are islands such as Malta, Barbados, Hong Kong, Kiribati, and Singapore. He believes that since only 10 percent of

---

2. John Gillis (2007, 281) cites the work of other scholars in his claim that European development from the Middle Ages was based upon the concept of the island. He describes a spiritual landscape of isolated churches, monasteries, pilgrimages sites, and so on linked together into a network that he labels as archipelagic.

3. DeLoughrey (2007, 12) speaks of the Western construct of island space as "islandism," a form similar to Edward Said's notion of Orientalism.

the world's population lives on islands, this distortion of island space can easily endure (203). The "mythic geographies" (Gillis 2007, 281) and ideological landscapes (Rose 1983, 87) that construct islands differ from the actual geographical diversity that exists. Conceiving of islands as small, remote, and abandoned represents a social construction of space. Further, given that islands are also represented as easily conquered, tamable (Lowenthal 2007, 206), available (Baldacchino 2007, 166), and therefore feminized (Addison 1995, 687; Rose 1983, 57) represents a masculinist construction. Edward Soja (1989, 79) observes, "Space in itself may be primordially given, but the organization, and meaning of space is a product of social translation, transformation and experience." In this regard, Soja prefers to speak of "spatiality" as a means of transcending the physicalist overtones in the use of the term "space."

Yet, even as we acknowledge the social construction in the representations of island space, we cannot avoid paying attention to space. The unique geographies of island spaces require consideration of how those spaces shape the mind. Unlike continents whose landscapes have largely been subdued in order to facilitate social activities, island spaces remain, not to give any support to the standard tourism brochures, untamed and untamable in some respect. Therefore, spatiality, as used by Soja, provides only a partial window for understanding island spaces. Karen Fog Olwig's (2007, 261) sense of the island as transspatial, creating opportunities for opening to the world, adds to this discussion. Olwig offers the term "global islandscape" to pay attention to islands socioculturally. And while she focuses on the portability of the islandscape, we find the need to give consideration to the physical environments that give shape to the islandscape. Philip Conkling (2007, 199), in defining "islandness" as a mental construct, derives his point of departure from the geographies of islands: "the rhythms of tides, wind, and storms [that] determine what you do and will not do." Although we may not follow Conkling in depicting the relationship between island and resident as one of sheer determinism, his assertion that the island geography, marked by isolation, shapes "islandness" proves useful for our purpose: "We think of islandness as a metaphysical sensation that derives from the heightened experiences that accompany the physical isolation of island life" (200).

Island spaces produce different mental impacts that do not all emanate from long-term residence on islands. Island spaces also shape the minds and imaginations of those who do not reside on islands. John Gillis (2007, 274) offers the description of "islomania," a term he derives from

Lawrence Durrell's *Reflections on a Marine Venus*, as a mental condition where people find islands enticing. The island serves as the place to encounter enchantment and mystery, to live out dreams and work through nightmares. This fascination with islands, Gillis notes, extends to several areas of modern life even in technological language of "surfing" and "navigating" (276). Another mental condition worth noting is "nesomania," which DeLoughrey (2007, 6) describes as "obsession with islands [as] a main feature of European will to empire." Operating both as "objects of desire"[4] (Garuba 2001, 61) and strategic possessions in the expansion of imperial maps, islands convey additional resources, coastlines, potential military bases, and economic assets to empires. Jon Heggulund (2012, 112) develops the thesis proposed by Halford Mackinder that Britain excels as an empire precisely because it extends its territory into maritime space, thereby being "at once bounded *and* extended by the sea." These two mental impacts account for the fantasy of the island as tourist destination, ideal for dreams as well as the hegemonic hold on certain islands such as Britain's relationship with the Falklands (Malvinas) or the United States' continued hold over American Samoa or Guam. And in the Pacific Ocean, France is not ready to let go of Tahiti and New Caledonia while, further north, China and Japan dispute over the Senkoku Islands. The fantasy and possessive impulses value island spaces in ways that make them desirable to dominant cultures with the resources to either purchase these spaces outright or to "time-share" them.

Our examination of the link between thinking and island space requires that we conduct this investigation from the perspective of islanders. The history of exploration and imperialism conveys a high value to island spaces that persist in the modern socioeconomic and military constructs. These valuations, though, largely serve exploitative interests and underlie the ideological representations of island spaces written from the continental or non-islander perspective. Harry Garuba (2001, 66) makes the case for using "the island itself as the 'site for thinking.'" Responding

---

4. The easy transfer of islands as birthday presents or trades at the end of war in the history of modern imperialism represents Garuba's (2001, 61) idea of "the movement from exploration to exploitation" that marks the change in the function of islands from fantasy places to possessions. Larry Ellison's purchase of the Hawaiian island of Lanai represents a more contemporary case of the combination of fantasy and possession. Julian Guthrie reports on this purchase with the headlines, "Larry Ellison's Fantasy Island" (*Wall Street Journal*, June 13, 2013).

to what he calls "the narrative of the island," he shows that "island narratives" merely theorize the conceptions of islands in dominant discourses. That is to say, discourse about islands from the perspective of islanders amounts to simply writing back and critiquing the empire (64–66). Essentially, narratives of the island only feature the island and the island experience. Rather than representing an indigenous discourse, these narratives emerge from a view outside of the island and in the process produce reactionary discourses from the island. Escaping the block to creativity and generativity imposed upon island thought by imperialist cultures requires refocusing on the spaces of islands and their unique geographies.

One possible way to think through the island spaces is by the embrace of the margins. If islands spaces are construed as those pieces of land detached from larger territories, then rather than being seen as a deficit, this reality serves as a decided strength of island spaces, what Yi-Fu Tuan (1995, 229) regards as "both fate and a source of pride." Focusing on detachment, not in exoticized, romanticized, or exploitative ways, builds on the essence of marginality as bell hooks (1984, preface) offers: "To be in the margin is to be part of the whole but outside the main body." Island spaces occupy a unique position of strength through their marginality: they are both inside and outside of the continental spaces. Hooks (1990, 150) avoids thinking of the margins as a site of despair, lest "a deep nihilism penetrates in a destructive way the very ground of our being." This viewpoint embraces the margins as the site of productivity rather than a space of lack. By recasting the power differentials in ways that restate the power inherent in the margins, hooks offers a way out of the dilemma of constantly writing back or responding to the center. The embrace of the margins, hooks believes, enables the creative and resurgent work to take place in a space dedicated to productivity and generativity (152).

In the geography of the modern world, islands occupy the peripheries of built-up areas.[5] Single page global maps omit most islands, thereby visually inscribing their marginal locations.[6] Godfrey Baldacchino (2007,

---

5. The exceptions to this rule being islands like Manhattan and Singapore Island (Pulau Ujong) that form the core of urban centers and are connected to major territories by bridges and tunnels.

6. Stephen Wright reminded us that the dominant mapping system—the Mercator projection—inflates the size of land as you move further from the equator. So Europe looks larger than it actually is, while pacific islands close to the equator, by contrast, look small and unimportant. The Gall-Peters projection was developed to

166) describes the peripherality of islands as "being on the edge, being out of sight and so out of mind." The benefit of this disconnectivity lies in what he views as the malleability of islands, even a "threatening fluidity" (Heggulund 2012, 111, with reference to Brathwaite 1983). Baldacchino formulates his thinking under the influence of Edward Kamau Brathwaite's idea of the peripheral island that occupies a disruptive space in the geographies and histories of empires. Brathwaite represents this disruption in his understanding of islands as offering alter/native discourses. He writes, "The alter/native. Not native. Note. Not simply native. Note. Natives are too easily exterminated as you know" (1983, 35). The alter/native conveys the peripheral space that islands occupy and from that space exercises a disturbing geographic presence that at once breaks up the monotony of oceans, thereby offering strategic and economic refuge to the adventurous but also disappearing off of maps, belying expectations of permanence or stability. More than simply foregrounding island spaces as ambiguous, the notion of peripherality and its correlate of alter/native present the opportunity of seeing island spaces not simply as responses or write-backs to nonislands spaces, but rather as spaces of originality and innovation.

Islands admit innovation in ways that make island spaces at once dependent and fiercely independent. Rethinking island spaces as more than simply land and paying attention to the surrounding waters enables the conception of the geography of island spaces to be what DeLoughrey (2007, 2) calls "terraqueous." And while she deploys the term to describe the globe and thereby render all landforms into islands, her understanding of the seascape as a critical part of island space presents the opportunity for decentering power. The sea enables the undoing of the negative consequences of territorial conquests and opens avenues for charting new paths. Islands, precisely because of their proximity and interaction with the sea, enable this decentering in unique ways. DeLoughrey locates the innovation of terraqueous space in the rewriting of history through the scripting of previous marine histories of islands in the face of hegemonic colonial histories (21). But we see even more innovation as islands negotiate their space in what Havea in this anthology regards as their liquid existence. Bounded and contained by the sea, islands not only make the "perfect prisons" but that same isolation grants autonomy to islands, metaphori-

---

give a more accurate depiction of relative land size, but it has not been adopted widely. Everyone, it seems, is used to America and Europe appearing bigger than they are.

cally producing in Lowenthal's term the "*I*-land" (2007, 217–18). Circumscribed by the sea, the I-land avoids the self-centeredness of acquisitive territoriality, the egotism that breeds the chauvinism of race, nationality, creed, and so on. The *I*-land may assert individuality as a result of its separateness, as Tuan (1995, 229) observes, but what Conkling (2007, 200) refers to as the "obstinate individuality" that marks islandness remains also "highly communal." Precisely because geographically islands respond to their place in the sea, island spaces stand at once closed as much as they are open. John Donne only partially understands this in his poem "No Man Is an Island," which undercuts the claims to absolute individuality. But his sense that the "I" of British modernity cannot be compared with the island/*I*-land, since islands are connected to continents, misunderstands the geography. Islands exist not simply as "a piece of the continent," but as parts of the sea.

The geography of islands requires that we pay attention to spaces of land and sea. This interaction produces what Brathwaite calls tidalectics (1983, 42), a way in which islands navigate their relationship with the sea. Brathwaite's neologism places the emphasis on the sea (the tidalectics of the sea), since he understands that "the sea influences the landscape" (Brown 2004, xiii). His point lies more in the simple lessons of tide actions and even more than the bare metaphor of the tides. Brathwaite's tidalectics serve as an organizing tool for thinking of the varied histories that mark island existences in the modern world. Even as we engage histories and theories, the geographies of island spaces compel consideration.

RUM-I-NATIONS

Shifting to the "I" in rumInations, we come to "islandness" or "islandedness." What might it mean to read in an island-infused, island-informed, perhaps even "insular" (read both for its negative and positive potentials) way? This question is at the heart of these essays and informs many of the ruminations that have led to this collection. Underlying these is the fundamental question: does "place" matter for interpretation, and if it does, how does it matter? The purpose of these essays is not so much to solve this problem as to raise the question.

The ruminations that come together in this collection follow various tidal currents, sometimes slapping into one another like Havea's *talanoa* (in this volume) and sometimes speaking in dialects all their own, like the Gallic languages of the Hebrides. Several address the closest geographic islands to the North American mainland: the Caribbean archipelago. Here questions regarding liberation, creation, and identity commingle in the creolization of island life, bounded and, paradoxically, linked by the sea. Other essays address primarily the islands and islanders of the Pacific. These raise questions about biblical interpretation, biblical translation, and the telling of tales as these reflect Pacific Island cultures. As readers, you are invited to follow these watercourses, to venture onto these different interpretative "landings." You are invited to consider with us what, if anything, might be that "insular" perspective that characterizes these differing readings.

Among the questions that emerge, perhaps the first is "What is the characteristic of an island?" Are islands connected or separated or both? Are islands defined by their isolation and thus by what happens on the land? Do their boundaries define them, if one can think of the sea as a boundary? Or, are islands defined by that space in-between, that ring of sand that stands as a metaphor for that place of creolization, of land-meeting-sea-meeting-land, that commingling of elements that ultimately all landmasses share, but not to the same extent as islands? Is the insular quality of island sensibility governed by the relatively high ratio of boundary to place, by the unusual amount of interstitial space that must have some impact on how "place," and thus identity within place, is understood?

Certainly this more extensive sense of interstitial space, of boundaries that define who "we" are, must have something to do with what it means to think as an island person, to interpret in an "islanded" way. Whether the island sits in the midst of an oceanic archipelago connected by liquid highways or next to a continental mass connected by bridges and tunnels, there is still the sense of "we," of insular identity, created by the presence of physical boundary in every direction that is somehow different than one might feel in the middle of a continental landmass. But then what? How does a heightened awareness of extensive boundaries, and/or of connection by human-manufactured means—whether bridge or boat—affect interpretation of world, of self, of place, and, for our purposes, of texts and contexts? How do we, who may see interstices both as constraints to intercourse and as invitations to different modalities of interaction, see the interstices of textual aporia, of canonical order and textual variant, of narrative silences and theological disagreements? And facing these boundaries, what might our insularity teach us to do at these interstitial places, places of identity and connection, of invitation to self-definition and to bridge- and boat-building? And given these insular instincts, what might be the benefits and costs of such activities, both for ourselves and for the (de)constructive study of biblical texts?

The essays in this volume highlight that one of the clear connections between our insular readings and other readings on landmasses are ongoing concerns about imperialism and its corollaries of (post)colonialism, the exploitation of scare resources, and the importation and/or resistance toward culture and material from "off-island." These consonant ideas are certainly used and appreciated, as are several goods imported to islands. But the presence and (perhaps uncritical) use of (primarily continental) critical theory raises a caution also, for the presence and ubiquity of this theory within our discourse points to the permeability of those interstitial spaces that define what is "island" from what is "not island." Boundaries, after all, are not only limiting, they can also be protective. And there is always a potential danger to that which comes to the island by boat—or on the currents of the air or by surfing radio waves. While creolization is, as Miller argues in this volume, an identifying feature of the insular subaltern, at what point does creolization give way to colonization? At what point does that which comes across the waves so reinvent the island after the culture of the mainland that the island itself no longer exists? And to what extent do "mainlanded" theories threaten to drown out other sorts of questions that derive from the peculiarities of insular life? Is there a theoretical response to (post)colonial, liberation, and other such theories that is particular to island readings? And if not, how might these insular readings escape being swallowed up by the larger intellectual continents surrounding them, reduced to dots on the intellectual map, barely visible? How do they escape becoming tourist destinations, full of exotic stereotypes to be seen and exploited, and then encapsulated by some native-made trinket mass-produced on a landmass, imported on the island, sold to the tourists and set on a pedestal to demonstrate the worldliness that is a soft form of conquest?

The corollary to these questions is about biblical interpretation. This volume is an attempt not only to think theoretically about islandness, about insularity made concrete as well as conceptual, but also to think and theorize about the biblical writings as insular, islanded writings. It is an attempt to read from the island back to the mainland, to follow the currents that surround the always shifting shoreline for the purpose of reading Bible and sometimes also reading scriptures.[7]

---

7. Wilfred Smith (2005) makes the same distinction, namely, that scripture is a

So, these questions about insularity, about the differences between and consonances among insular and other readings of texts (postcolonial, liberation, etc.), undergird another set of questions about biblical interpretation. How can questions about the nature of insularity be used to think about biblical writings, communities, and formations? What, if anything, is insular about biblical communities, whether under imperial or parochial monarchy, in exile or illicit? In what ways are biblical writings like islands? And if they are like islands, how are these islands related and/or relatable? Are there set routes between the islands, or do we, as insular readers, have the freedom to negotiate those interstices as the tides of interpretation may take us? Are these "islands" separated or connected, by what human inventions, and in the face of what sorts of dangers? (After all, bridges collapse and boats sink). What might the connections between these insular texts negotiated through (perhaps dangerous) waters entail? What, if anything, takes place when these stories, like *talanoa*, slap together? And if the result is violence, is that violence destructive and life crushing, like a slave master appropriating Luke 12:47 to justify physical brutalization in on-island slavery? Or might that violence be both destructive and creative, like the creation of island mass from the explosion and expulsion of hot lava in volcanic eruption?

Aymer also raises in her contribution the question of the consonance of the foregoing questions with others who study islands. For we are not alone in raising these questions of what it might mean to think about insularity as a physical, geographic reality. There is an entire discourse forming among geographers, sociologists, and historians that attempts to establish the contours of geographic insularity. Among these writers, the study they are attempting is called "island studies." Geographer Pete Hay writes in the first issue of *Island Studies Journal*, "The metaphoric deployment of 'island' is, in fact, so enduring, all-pervading and commonplace that a case could reasonably be made for it as the central metaphor within western discourse" (Hay 2006, 30). However, these metaphoric descriptions strike Hay as not the subject of islander studies at all. He writes, "I do not believe that they fit within the purview of nissological investigation, which should, rather, concern itself with the *reality* of islands and how it is for islands and

---

human activity, that human beings and human communities designate and treat texts as scriptures, and that no writing, not even a biblical writing, is ontologically scripture.

islanders in the times that are here and that are emerging" (Hay 2006, 30, discussed by Aymer in this volume).

Hay's challenge raises a concern as the question of insular readings of biblical texts continues and hopefully expands. If we hold that one of the permeable, insular boundaries that are fundamental to island studies is not the metaphor but rather the reality for islands and islanders, what questions are we forced to ask and will these be questions that can—or do— intersect with biblical interpretation? Perhaps, it will invite more explorations of popular readings of biblical texts by island peoples, as is evident in the essays in this volume by Macaskill, Ma'ilo, and Vaka'uta. Perhaps it will encourage more historical consideration of the role of islands in ancient biblical life, as does Kinukawa's essay in this volume. And surely, as Middleton notes in his essay, those readings will cause other intersections, intersections not only with postcolonial and liberationist concerns but, in very specific ways, with ecological readings as the planet warms and the survival of many islands are threatened with rising seas.

A topic unaddressed in this volume is the question of the negative connotation of islands. For there are other islands not represented in this volume that may well prove fruitful discursive places for biblical interpretation in the future. We are thinking here of islands like Robben, Elba, and Alcatraz, islands of exile whose boundaries serve not as passageways but as one of several prison walls keeping their islanders trapped. Here, too, we would put islands like Angel and Ellis, islands as interstices, protecting but never part of the continent nearest to them. Here the term "barrier island" is perhaps useful; what might this mean and how might this affect biblical interpretation? To this list, we would add islands of ambivalence, whether because of a "positive" exoticization (e.g., Hawaii to the United States or the French Antilles to France) or because of a concerted effort of vilification (e.g., the relationships between Cuba and the United States or between Taiwan and China). In addition, we would add imperial islands, not the least of which might be Japan (see Kinukawa in this volume) and the British Isles. As we continue thinking together about islands and biblical interpretation, what might we learn by taking seriously these islands also? How might they contribute to a broader understanding of "island" reading?

Finally, we return to the question of theory. What might it mean to theorize about islands? How might we take Vaku'ata's critique of mainland theories of ecology and postcoloniality, Spencer's creolization, and many of the other theoretical sketches seriously? Is there such a thing as "an" island way, "an" insular way of theorizing about islanded interpretations? Or might it be the case that islanded interpretations themselves exist as islands, connected and creolized but separate and speaking in the kinds of diverse voices and vocabularies that island peoples often use, even those who share the same geographic region and language? And if this latter, what might it mean to bring our Bab(b)el of voices together in the spirit of *fale-'o-kāinga* (Vaka'uta in this volume) acknowledging that we islanders and former islanders theorize our biblical interpretations both individually and reciprocally, with respect for each other? How might such an island-centered way of reading together affect even mainland discourses in the future?

rumiNations

We shift again, this time to the last part of the subtitle: rumi-nations. How might we island-think about the nation thing? How might island space and islandness help us rethink nation, nationalism, and nationhood? This section involves turning (without departing) from island space and islandness (identity) on to island-thinking, bearing in mind that these strands interweave: thinking is formed and conditioned in and by space and by who we are (identity). There will therefore be some rewinding and fast-forwarding between island space, islandness, and island-thinking, as we hone in to "nation."

Insofar as "island" is not an automatic cue for "nation," nor does "nation" cue "island," we attempt in this section to island-think something that is not usually associated with islands. We hereby address a blind spot in the usual conceiving of islands, namely, that islands are not nations. Our attention shifts to the nation thing, but we are still very much in island mode. In this regard, this section engages in out-land-ish reimagining of the hermeneutics thing.

There are many connotations of the word "nation," but they intersect around the notion that a nation is made up of a group of people who come together because they share certain things. Their coming together may have been accidental in the first instance, but after some time, over several generations, they would learn to gel. What is shared varies from nation to nation, depending on the heteronormativity of each nation (Spivak 2009): it may be a combination of common ancestral roots, heritages, beliefs, cultures, languages, ethnicity, lore, government, territory, and so on. A nation is thus not much different from an island, which is also made up of people who share common things that distinguish one group of nationals/islanders from others.[8]

Nations come in various sizes and colors. Some are broad, tall, and heavy; some are narrow, shallow, and physically challenged; and some nations are like "periods"[9] that drift in the sea. Some nations are bright and flashy, some exude warm tones, some are dull, and some are repelling. Size and color do not explain why and how nations come into being or their power and wealth. Put another way, the weight and influence of a nation does not depend on its terrains. Many nations are islands and archipelagoes; many islands are nations. Nations they all are, no matter their color and size.

Nations are born from the adherence of a community of peoples to a combination of common things. This does not mean that all nationals understand and value those common things in the same way or have the same list of common things. They do not need to be of the same mind, but they need to feel that they belong to the common things that were vital to the birthing/berthing of their nation. A nation is not fixed to a specific place and time but, like a wave in the sea, rolls out in response to their conditions. The devotion of the community (nationalism) and the identity they take on (nationhood) contribute to defining the power of their nation. In this way, small (island) nations can and some do have a lot of influence and power. Moreover, there is a better chance for nationals of smaller (island) nations to intimately know the common things around which they are formed than members of larger nations do. In fact, the smaller the nation the more tribal its nationals tend to be, or put differently, insular, islanded.

In Gayatri Spivak's (2010, 13, 21) estimation, agreeing with Eric Hobsbawm, "there is no nation before nationalism" and "imagination feeds nationalism." There is public-private crossover here insofar as nation is a public entity while imagination kindles in the private realm. Nationalism is a condition for the birth of a nation, and a nation grows

---

8. Several nations may have the same structures and set up, but differ in confederation. An example of this is in 1 Sam 8. The people asked for a king to govern them "like all other nations" (8:5), but at the same time they wish to remain separate from those other nations. The kingship will make them both *like* and *different* in relation to other nations.

9. This is a popular image among islanders in Oceania, referring to their islands as periods or dots (like full stops on paper) in the sea. The image is pregnant with meanings.

and matures when its borders are drawn and secured. There is no nation without borders, which nationalists protect as lines for exclusion. Like religions, nations are born in response to zeal and yearnings that draw people together. Religiosity and nationalism are sources of legitimacy (see Spivak 2009, 78), seasoned with the smugness of tribalism, and they help people belong across borders (see also Havea, Neville, and Wainwright 2014).

The 1947 partition of Pakistan from India involved establishing borders between the two republics. Pakistan became a nation when borders were drawn between it and India,[10] similar to the Demilitarized Zone that separates North Korea from South Korea (recognized in 1948 and one of the points of conflict in the 1950–1953 Korean War) and the fence that divides the United States from Mexico. Nations are more than their borders, but the establishment of borders is necessary. This stance raises questions about the two-state solution proposed for Israel and Palestine, according to which Palestine is to be like periods or islands within the borders of Israel. Dotted in various places in Israel, Palestine is to be like an archipelagic island nation. The late Edward Said (1999) rejected the two-state solution on the grounds that it would foster apartheid. As a nation dispersed within another nation, Palestine is to be a network of communities, like a sea of ghettoes, rather than a sovereign body. The communities of Palestine can still share certain things in common, but their borders will not be connected. What nation will Palestine be if it, though driven by religious and nationalist motivations, does not have its own borders? In the end, the two-state solution will raise a "security fence" for some and a "separation wall" for the rest (Chomsky 2007, 29–34, 63–66).

Borders exist because of what lie outside of them. There is no border if there is no outside; there is no nation if there is no other group/nation(s). If there is no outside, there is no inside and no borders either. Can there be nationalism without borders? We accordingly propose to supplement Spivak's assertion: there is no nationalism without borders and outside(r)s, real and imagined. We make this proposal as islanders, because borders and outside(r)s are constant in the minds of islanders. Because island space is limited, islanders can see, hear, and smell island borders every day and gaze "outside" on to the horizon and yonder. Islanders are border peoples, similar to Rahab who lived at the outer wall of Jericho (Josh 2:15;

---

10. A different process was involved in the 1971 partition of Bangladesh from Pakistan, because India was already in between those two republics.

see Vaka'uta in this volume), and similar also to the Midianites, Amalekites, Moabites, and other wilderness peoples whose homes are in the Bible's hinterland. Border peoples are alert to those who cross inside and outside. For them, borders are not lines of separation or partition but places of dwelling and places for engagement. Borders are not barriers but places of intersection and of negotiation, of going and coming, of transiting and emigrating. With this perspective, a simple equation is drawn: as islands are nations so are borders "borderlands" (borrowing from Anzaldúa 1987).[11] Whereas Anzaldúa's Chicanas/Chicanos and mestizos experience borders as barriers, islanders relate to the border (ocean) as their home.[12]

The borderness of islands and islanders invites an alter/native way of imagining nationhood. We thus propose another shift, from seeing nations as entities surrounded and defined by borders to seeing nations as borders. Nations are borders not in the exclusionary way experienced by Chicanas/Chicanos and mestizos at the United States border, but in the homing way that the ocean is to islanders.

> The skin of the earth is seamless.
> The sea cannot be fenced,
> *El mar* does not stop at borders. (Anzaldúa 1987, 3)

Nations are not permanent destinations but places for crossing and intersecting, for transiting and negotiating. Nations, no matter their size and color, are stepping-stones that point and lead away from themselves. Nations are born because of peoples and their spirits of nationalism (tribalism), and as borders they exist because of and for outside nations. As borders, nations exist because there are other nations outside themselves. Without outside nations, they cease to be nations and borders lose their borderlines.

The borderness of islands invites revision of our understanding of nations. Nations are not bodies that are distinguishable and separate from

---

11. Borders push the Chicanas/Chicanos and mestizos back. They are not to cross over, as gringos freely do. Chicanas/Chicanos and mestizos are condemned to the borders, which has become the home for many, hence the notion of borderlands. This is why *polleros* (coyotes) are critical to help people across the borders into the United States so that they might find work (see Smith-Christopher 2007, xvii–xxi).

12. Relevant for this rumination is how Anzaldúa (1987, 1–3) began by talking about the ocean, which she distinguishes from the fences that divide landlocked nations.

one another only, but collectives that are in relation to one another. No nation *is* on its own. Ubuntu! In other words, to borrow from the Vietnamese Buddhist activist Thich Nhat Hanh, nations "inter-are":

> You are me, and I am you.
> Isn't it obvious that we "inter-are"?
> You cultivate the flower in yourself,
> so that I will be beautiful.
> I transform the garbage in myself,
> so that you will not have to suffer.
> (Hanh, "Interrelationship")[13]

The ocean links *I*-lands up. While there are wide distances between island nations, spread out like periods in the sea, islanders do not grieve over our separation as something that impoverishes island living. We do not deny that many islands float in isolation, but we argue that isolation is not a threat to islanders as much as it is for nonislanders. What is isolation to people who are isolated? As people who live in water do not know what it means to be wet, islanders who are isolated from everybody else do not see isolation as a problem. There is another explanation for this untroubled mind-set: Island worldviews are not landlocked,[14] so distance and separation do not automatically add up to isolation. This paradoxical position is evident in the Tongan saying *'auhia kae kisu atu pē* ("drifting away, but reaching to you"). It is possible in the island worldview to be distanced (in space) and at the same time be connected (in relations). In other words, islanders are relational people, and isolation has to do with relations rather than with distance.

Relations are woven in the interaction between people, obliging one to another, and people are attached to some places because of the relations that those places call to mind and represent. Islanders attach to island roots and island homes, because the islands "contain" our ancestors, heritages, and customs. We might drift away over the seas to faraway lands, but we can maintain our relations and thus continue to "be in touch." Ones who move away are not in isolation, nor are the ones who remain at home

---

13. For the complete poem, see http://allspirit.co.uk/interrelationship/.
14. The fence is menacing to Anzaldúa's Chicanas/Chicanos and mestizos, but the ocean is inviting to islanders.

islands. So we define our relations not only by where we are (place and distance) but the connections (ties and relations) that we maintain.

In the eyes of many nonislanders, islands are outside their national borders so islanders are therefore outsiders. The borderness of islands, to the contrary, imagine nations in relation to one another. As border peoples, islanders are relational peoples. The interweaving of borderness and relationality conditions the worldviews of islanders in ways that are different from peoples in landlocked nations, who are separated from those on the other side of the border.

Landlockedness is difficult for islanders to comprehend. When a border divides an island up, the border was introduced and is maintained by some colonial force. The odd instance of France and Holland dividing up the thirty-four square mile island of Saint Martin/Sint Maarten is a case in point. In the case of Papua New Guinea and West Papua, the colonial force is Indonesia. Colonialism has thus introduced landlocked borders to island settings. We put Spivak's argument to the lines of island-thinking: there is no nation before nationalism, and nationalism often falls at the feet of colonialism. Colonialism continues to boil the sea of islands in Oceania, with France, the United States, and Indonesia holding fast to some island groups. Colonialism erects borders and extinguishes the spirit of nationalism, experienced most severely by smaller island states.

This section applied island-thinking to the complex relation between nation, nationhood, and nationalism, with scriptural interpretation lurking on the shoreline. How might this island-think on the nation thing help form the island hermeneutics thing? There are several pathways.

First, insofar as scriptures contribute toward sparking nationalism within and in front of texts, it is vital to check the temperature of both nationalist texts and readers. Whose imaginations and interests do they manifest? What kind of borders do they erect? Against and/or for which colonial force do they stand?

Second, insofar as scriptures have been used to blow wind on the sails of colonialism and continue to be used in that way in the modern period, then island hermeneutics joins arms with other hermeneutics of suspicion and of resistance in advocating minority (islandish) and minoritized subjects and interests (Bailey, Liew, and Segovia 2009). The nation thing infers that texts and readings are driven and so, by transference, no reader should be a bystander. This has to do with the relational island thing, whereby one may drift away yet reaching out (*'auhia kae kisu atu pē*). Relations begin between individuals and then extend toward families and communities. Along this line, extended family, rather than nuclear family, is the island thing.

Third, insofar as the leanings of border and relational peoples are stronger toward cooperating than toward conquering, dividing, and dispossessing, collectivity is an apt goal for islandish readers. The nation thing of bringing peoples together is also an island thing. The island thing is not just in response to the call for regionalism (see Spivak 2009, 88) in order to break through the borders of nationalism but in favor of islandedness. In this way, islands and islanders from different regions and oceans may form a collective. Herein is a chance to propose "equivalence":[15] texts and readings from different textual regions may be drawn into a collection. This is one of the reasons behind this collection of essays and of the new name of the Society of Biblical Literature group—Islands, Islanders, and Scriptures—that b(i)erthed this volume.

iSlands and iSlanders

On the tongues of creolization (see Miller in this volume), we close by putting the markers of our landings of who we are, islands and islanders, as it were, upon the twangs of rumination. There are particular but not unique slants ("slands" in islands), skews of islands and islanders—around the intersection of island space, islandness, islandhood, and Bible—presented in the leaves of this anthology. Some islanders will be nauseous because of those, and we imagine that some nonislanders might want to go down those slants. Whether to opt out or to be hooked up, we will not be put out, for after all, exit and entry, arrival and departure (see Davidson in this volume), are island slants also.

The slanders, by islanders and nonislanders both, against the island things and island-thinking do trouble us. This anthology hopes to bring those islanders into the course of biblical hermeneutics. For the sake of kindling some relief, we rewind to our opening proposition: biblical texts are like islands, and readers are like islanders.

---

15. The appeal here is to Spivak's (2009, 81) claim that equivalence is the stuff of orality (an island thing also): "If the main thing about narrative is sequence, the main thing about the oral-formulaic is equivalence."

Works Cited

Addison, Catherine. 1995. "'Elysian and Effeminate': Byron's *The Island* as a Revisionary Text." *Studies in English Literature, 1500–1900* 35:687–706.

Anzaldúa, Gloria E. 1987. *Borderlands/La Frontera: The New Mestiza*. San Francisco: Aunt Lute.

Bailey, Randall C., Tat-Siong Benny Liew, and Fernando F. Segovia, eds. 2009. *They Were All Together in One Place? Toward Minority Biblical Criticism*. Atlanta: Society of Biblical Literature.

Baldacchino, Godfrey. 2007. "Islands as Novelty Sites." *Geographical Review* 97:165–74.

Brathwaite, Kamau. 1983. "Caribbean Culture: Two Paradigms." Pages 9–54 in *Missile and Capsule*. Edited by Jürgen Martini. Bremen, DE: University of Bremen.

Brown, Stewart. 2004. Introduction to *Words Need Love Too*, by Kamau Brathwaithe. Cambridge: Salt.

Chomsky, Noam. 2007. *Interventions*. Camberwell, VIC: Penguin Books.

Conkling, Philip. 2007. "On Islanders and Islandness." *Geographical Review* 97:191–201.

DeLoughrey, Elizabeth M. 2007. *Routes and Roots: Navigating Caribbean and Pacific Island Literatures*. Honolulu: University of Hawaii Press.

Fewell, Danna N, ed. 1992. *Reading between Texts: Intertextuality and the Hebrew Bible*. Louisville: Westminster John Knox.

Garuba, Harry. 2001. "'The Island Writes Back': Discourse/Power and Marginality in Wole Soyinka's 'The Swam Dwellers,' Derek Walcott's 'The Sea at Dauphin,' and Athol Fugard's 'The Island.'" *Research in African Literatures* 32:61–76.

Gillis, John T. 2007. "Island Sojourns." *Geographical Review* 97:274–87.

Havea, Jione, David J. Neville, and Elaine M. Wainwright, eds. 2014. *Bible, Borders, Belongings: Engaging readings from Oceania*. SemeiaSt. Atlanta: Society of Biblical Literature.

Hay, Pete. 2006. "A Phenomenology of Islands." *Island Studies Journal* 1:19–42.

Heggulund, Jon. 2012. *World Views: Metageographies of Modernist Fiction*. New York: Oxford University Press.

hooks, bell. 1984. *Feminist Theory: From Margin to Center*. Boston: South End.

———. 1990. *Yearning: Race, Gender, and Cultural Politics.* Boston: South End.
Lowenthal, David. 2007. "Islands, Lovers, and Others." *Geographical Review* 97:202–29.
Olwig, Karen Fog. 2007. "Islands as Places of Being and Belonging." *Geographical Review* 97:260–73.
Rose, Gillian. 1983. *Feminism and Geography: The Limits of Geographical Knowledge.* Minneapolis: University of Minnesota Press.
Said, Edward. 1999. "Setting the Record Straight: Edward Said Confronts His Future, His Past, and His Critics' Accusations." Interview by Harvey Blume. *The Atlantic Online*, September 22. http://www.theatlantic.com/past/docs/unbound/interviews/ba990922.htm.
Smith, Wilfred Cantwell. 2005. *What is Scripture?* Minneapolis: Fortress.
Smith-Christopher, Daniel. 2007. *Jonah, Jesus, and Other Good Coyotes: Speaking Peace to Power in the Bible.* Nashville: Abingdon.
Soja, Edward W. 1989. *Postmodern Geographies: The Reassertion of Space in Critical Social Theory.* London: Verso.
Spivak, Gayatri Chakravorty. 2009. "Nationalism and the Imagination." *Lectora* 15:75–98.
———. 2010. *Nationalism and the imagination.* Calcutta: Seagull.
Sugirtharajah, R. S. 2003. *Postcolonial Reconfigurations: An Alternative Way of Reading the Bible and Doing Theology.* Saint Louis, MO: Chalice.
Tuan, Yi-Fu. 1995. "Island Selves: Human Disconnectedness in a World of Interdependence." *Geographical Review* 85:229–39.

First Waves

Islandedness, Paul, and John of Patmos

Margaret Aymer

Back to Africa miss Mattie?
You no know what you da seh.
You have feh come from someplace fus
before you go back deh.
                —Louise Bennett-Coverly, "Back to Africa," 1994

Babylon, you throne gone down, gone down …
Fly away home, to Zion …
One bright morning when man work is over man will fly away home …
All you shall witness when Babylon fall.
                —Bob Marley, "Rastaman Chant," 1973

Two discourses. One island. Is there something about the "islandedness," to coin a word, of these discourses—the first from Louise Bennett-Coverly and the second from Robert Nesta Marley—that helps us to understand how both of these emerge from the same island? And, through that same lens of islandedness, can we then begin to unpack another set of oppositional discourses: the "grace and peace" of Paul of Tarsus and the "come out of Babylon" of John of Patmos? In this short essay, I sketch out some ruminations around this concept of islandedness and how it might function as a wedge with which the study of the Bible and the ancient world might be split open in new ways.

I come into this discourse by heritage more than by experience. I am a first-generation former islander, a migrant to the United States. My northeastern accent with slight British overtones learned at the feet of my anglophilic parents and my somewhat fluent code-switching into African American rhythms racialized upon me in the United States betray none of the cadences of my natal island, Barbados, and few of the cadences or languages of my islands and mainlands of family lineage: Jamaica,

Panama, India, Saint Lucia, Montserrat, Antigua, and Barbuda. The truth is, I spent almost as many years living the "islanded" life on Manhattan as I did living in the Greater and Lesser Antilles of the Caribbean. So, I come at this task without any illusions of essentialist "islandedness" by virtue of the 169 square miles of coral and limestone on which I drew my first breath or of the islands on which I learned that "m" was for mango and that "Chi Chi Bud Oh" had as its necessary response "Some a dem a halla, some a bawl."[1] Instead, I come at this reflection on islandedness by using islands, in the words of historian John R. Gillis (2004, 42), "to think with."

Toward the end of this essay, I will raise some problems I foresee for this conversation as a whole, and my stance as "island-thinker" will figure among them. However, I begin by sketching out the interplay between islandedness, migration, identity formation, and the choice of insularity, a choice that does not necessarily derive from islandedness. Then, I look at these same forces and choices at work in Galatians and in the Apocalypse of John, as well as implications of this mode of thinking for other biblical texts. As I close, in addition to the problems noted above, I address some ways in which this wedge of islandedness might open up different conversations for biblical studies research.

Roots and Routes

Godfrey Baldacchino (2004, 274), Canada Research Chair in Island Studies at the University of Prince Edward Island, suggests that islandedness revolves around twin intersecting phenomena: *roots* and *routes*. The second of these, routes, designates the constant, community-(re)shaping presence of migration. Islanders migrate, both because of the limitations of being surrounded by water and because of the possibilities that inhere in the constant presence of a natural source of transportation. Among those limitations, the finitude of resources, opportunities, and sometimes even of safety lead to intentional movement among islands, either through the impermanence of trade or the disruption of em/immigration. I am not, here, talking strictly about the movement that is drawn in

---

1. "Chi Chi Bud Oh" is a Jamaican work song that goes through the various birds of the island. The caller will call the sounds or names of various birds ("Some a grung dove"), and the chorus of laborers will respond "Some of dem a halla, some a bawl." It is perhaps one of the most commonly known folk songs of Jamaica.

particular directions by colonizing forces. Even before colonialism, one saw this phenomenon of migration among the indigenous peoples of the Caribbean. As geographer Pete Hay puts it,

> a state of peripatetic island-hopping is almost the island condition, and … this is an exuberant activity wherein "each new intruder finds a freedom it never had in its old environments."… In the mix of the old and the new, island identities shift—they are endlessly remade, but enough remains constant for the island to persist. (Hay 2006, 24)

In short, central to a consideration of islandedness as a metaphor is a consideration of the impact of migration on the heritage of islanders, that is, of "routes" on islanders' "roots." For, if the ever-present sea serves both as a border and a passageway, both a defining presence and an invitation to exploration, then the people shaped by both of these realities will have, as part of their cultural heritage, strategies of identity and group formation derived from the migratory nature of their cultures.

Of course, not all migration is equal. The migration for trade or commerce may include greater and lesser partners. Forced migration, as happened both through the Maafa and through the colonial importation of Asian workers from China and India, creates different senses of identity than voluntary migration. Economic migration may be temporary or permanent, and if the latter, the once bounded island may find its cultural heritage shaped not only by immigrants to the island but by emigrants from the island sending new cultural norms "back home."

All of this takes place in the midst of the active wrestling within the ever-migratory population with what it means to be "from here," making what "roots" might signify ambiguous. No self-evident answer is available when those "from here" might not be immediately identifiable by physical traits, common dress, or even common linguistic patterns. To further complicate matters, larger islands include multiple cultural norms, some established on water and migration and others on "inlandedness." And it is further complicated when what is "appropriate" or high culture becomes dictated by one small migrant group with a great deal of colonial or neocolonial power.

Thus, for islanders, routes shape the contours and cadences of roots. Sociologist Avtar Brah might call this phenomenon "diaspora space," defined as the place where multiple subject positions are juxtaposed, contested, proclaimed or disavowed; where the permitted and the prohibited perpetually interrogate; and where the accepted and the transgressive imperceptibly mingle even while these syncretic forms may be disclaimed in the name of purity and tradition. *Here, tradition is itself continually invented even as it may be hailed as originating from the mists of time.* (Brah 1996, 208, emphasis added)

I argue that, at least for the Caribbean basin islands situated in the midst of the seaways of the Atlantic, "diaspora space" helpfully describes the contested and continually (re)negotiated nature of what it means to be "from here."

Insularity and Other "Islanded" World Reactions

"Diaspora space" rather than "insularity" is the defining characteristic of islander identity formation. Insularity is not an *a priori* islander characteristic, but rather a choice in light of the contestation inherent in diaspora space. Insularity is "a concept created and manipulated strategically by the islanders themselves ... to establish and express social identity," as archaeologists Arie Boomer and Alistair Bright note. Boomer and Bright go on to assert,

> Such identities can engender in islanders the ideological strength needed to opt out or resist nation state incorporation.... Island populations, whether real or perceived, maintain an attitude of remoteness from their more imposing neighbours in order to preserve a distinct way of life: in essence their insularity. (Boomer and Bright 2007, 13)

Insularity is thus one response, but not the only or even the necessary response to islandedness. Insularity has, as a cognate, *marronnage*—that fleeing out of the world, creation of a new world, and (re)negotiation of that newly created world as posited by Vincent Wimbush (2004, 22–29) in *African Americans and the Bible*. In the ancient world, its cognate is the broad category of asceticism, particularly but not exclusively the eremitic desert mothers and fathers of North Africa.

However, there are other possible responses to islandedness besides insularity. Some of these may be derived from the work of cultural psychologist John Berry on the ways in which migrants create and interact with dominant culture. Berry's idealized types serve as broad categorizations of strategies of migration not only for migrants to large landmasses, but also within the migration-infused diaspora space of "islanded" life.

Berry proposes two variables when considering migrant cultural formations: migrant reactions to their "home" culture and migrant reactions to the "host" culture. Those who, upon migration, turn exclusively toward their home culture and reject the host culture he calls "alienated." The other extreme of this, the utter rejection of home culture and the exclusive turn toward host culture, he calls "assimilation." The rejection of both home and host culture, opting for a third way, Berry calls "marginalization." And, the attempt to create a new space informed positively both by home and host culture, Berry calls "accommodation" (2001, 616–21).[2]

Berry has a bias that derives from his clinical psychological practice of aiding migrants in their cultural adjustment to their new homeland. However, from an islander perspective, each of the world orientations that Berry describes is a survival strategy when living bounded on all sides by a sea that functions both as barrier and as superhighway, facilitating the constancy of migration into the contested diaspora space that is the island.

Insularity as popularly defined, that turning away from the "other," falls between two of these strategies, depending on the nature of the dispute. It may take the form of marginality initially—the rejection of both the home and host cultures. However, it can morph toward alienation as generations grow up in the new marginal reality and rename that marginalized, liminal reality "home" with the contested diaspora space functioning as "host."

Rastafari is perhaps the quintessential example from my heritage islands of insularity: marginalization at first and now alienation as the community has grown to international proportions. It is from this community that the lyrics of "Rastaman Chant" emerge. First released by Bob Marley in 1973, "Rastaman Chant" encodes the rejection of "home" culture (Jamaican governmental structures) and "host" culture (the colonialist intrusions of the United Kingdom and the United States) as Babylon. The chant expresses the desire to "fly away home" to Zion, which is figured in Rastafari as Ethiopia, the home of Emperor Haile Selassie. In the 1998 release *Chant Down Babylon*, a collection of Marley's songs reworked

---

2. Berry's migrant enculturation theory: (1) alienation: rejection of host culture but not home culture; (2) assimilation: rejection of home culture but not host culture; (3) marginalization: rejection of both home and host culture; and (4) accommodation: creative tension between home and host culture.

by contemporary singers and hip-hop artists, United States-born Busta Rhymes and Spliff Starr, both islanders' children, add to the rejection of "home/host culture" with the call to "clean out we self" and the promise that "all you shall witness when Babylon fall."

If we understand islandedness as that place of contested identity in the midst of migration, we can understand the response of Rastafari as the decision to develop separately, to maintain an identity that can be bounded more effectively than is possible if one relies solely on the saltwater boundary of Jamaica. And yet, the irony of this insular identity is that it draws heavily on the cultures it claims to reject. At the heart of the Rastaman chant, for instance, is a quote of an American hymn, Albert Brumley's "I'll Fly Away" (1932), no doubt brought over on those seas by missionaries from the United States. And the remix of Rastaman chant relies also on the migratory patterns of islanded people. For where else would Jamaican-American Busta Rhymes and Trinidadian-American Spliff Star learn to merge hip-hop, reggae, and Rastafari and to bring them back into the contested diaspora space of the "home" island(s) than in one of the largest noncontiguous British-Caribbean "parishes," Brooklyn, New York (a community that may even have more influence on the contested diaspora space of "home" than any group of migrants entering the island)?

In contrast, consider the poem "Back to Africa" by Louise Bennett-Coverly, the late poet-laureate of Jamaica—"Miss Lou," as she was known to the generations of Jamaican children raised by her on Saturday morning television. She both popularized and, through her ethnolinguistic study, justified the preservation and celebration of the *patois* of Jamaica,[3] the language that was systematically colonized out of my parent's generation and thus which I, like some other 1.5 generation children, can read and follow but not speak.

Bennett's preservation of a language polemicized as "bad" or "broken" English, as demonstrated in the poem "Back to Africa," might be seen as an insular move. But note how she uses that language to belittle the "Back to Africa" movement of 1960s Jamaica. She tells her constant foil, "Miss Mattie" that "you have to come *from* someplace fus before you go back deh." Then, she proceeds to rehearse for Miss Mattie the contested dia-

---

3. In 1991, L. Emilie Adams published *Understanding Jamaican Patois: An Introduction to Afro-Jamaican Grammar*. Since the turn of the millennium, the movement in Jamaica has been to call the Creole spoken by its people Jamaican English (see F. G. Cassidy and R. G. LaPage, *The Handbook of Jamaican English*).

sporic realities of islandedness: the migration and formation of the people by Africans, Jewish migrants, and the French (and she could have included many others).

Bennett's resolution to the problem of diaspora space is accommodation that even borders on assimilation: "Oono all bawn dun a bung grung. Oono all is Jamaican." I say borders because, while she verbally affirms the "Jamaicanness" of Miss Mattie's people, she signifies that affirmation in language that is not at all assimilationist. Jamaican *patois* represents an accommodation of the varieties of forced, voluntary, and imperialist migrants on the island of Jamaica and the creation of a new entity born out of the contested crucible of diaspora space. Miss Lou literally speaks in the language of cultural accommodation, a language, to paraphrase Brah, that is invented and hailed as being Jamaican from the beginnings of time.

It is possible to hear in this, and in the Jamaican motto, "Out of Many, One People" (an anglicization of the *e pluribus unum*), a tendency toward accommodating others without respect to culture. At the same time, the entire purpose of accommodation is the ability to belong in two worlds at once, that of the home and host cultures. As such, the creation of an identity—and even a language—that honors one's past while affirming one's full acceptance into the diaspora space is a fully accommodationist move. And certainly, Miss Lou's choice to state her case in Jamaican and not in British Standard English—in which she was also fluent—was a rejection of assimilation into British colonialism. Hay puts it this way,

> There may be times when a politics of place identity conjoins with an illiberal politics of exclusion, but that requires other factors to be potently at play. What is far more typical, especially where island identity is concerned, is that an identity politics emerges from a desire to defend difference against the totalising trends within globalization. (Hay 2006, 28)

Islandedness and the Bible

Using islandedness and the related conversations about diaspora space and migrant strategies of enculturation to "think" about the Bible yields a clearer understanding that much of the biblical library, even the New Testament, is a diaspora space, a collection of writings "in-migration"—writings written to, for, or by migrants, some forced, some voluntary. Let me sketch out two examples of how islandedness might be used as a wedge to open up new discussions of the New Testament.

Consider John, the seer on Patmos Island. Like his Rastafarian descendants and his Johannine cousins, this John issues a call to the seven churches to adopt a Christian cultural insularity, an attitude of *marronnage* from the Roman Empire, from Babylon. Throughout his "revelation," John's alienated stance is clear. Almost at the inception of the Apocalypse, he calls the cities of Asia Minor satanic. For example, he describes Pergamum as the location "where the throne of Satan is" (ὅπου ὁ θρόνος τοῦ σατανᾶ), a polemic probably aimed at the presence of the imperial cult and its requirement to make sacrifices to the emperor as a god (Rev 2:13). Likewise, he charges that some in Thyatira have learned "the depths of Satan" (τὰ βαθέα τοῦ σατανᾶ) (2:24). John's most pointed polemic takes place in Rev 13, the depiction of the two beasts. While it is customary for interpreters to focus on the larger of the two beasts, John's concern is also with the second, smaller beast. Note its description:

> And I saw another beast that was coming up out of the land, and it had two horns like a lamb and it was speaking as the serpent [or the dragon]. And *all* of the authority of the first beast, it exercised on the first beast's behalf. And it did [so] on the land and all of those sojourning in it, so that they shall worship the first beast, of whom the deathly wound had been healed. And it did great signs so that it might cause fire to come down out of heaven before the women and men. And it deceived those who were sojourning in the land by the signs which were given to it to do on behalf of the beast, while it said to those who were sojourning in the land to make an icon of the beast who had the sword's wound and was living. (Rev 13:11–15)[4]

With this second beast, John, represents to his assemblies the true nature of the "host culture," those who enforce the imperial cult in Asia Minor. John characterizes this "host culture" as one that speaks as the dragon (or serpent) itself. It has a satanic voice, although it mimics the Lamb in appearance. It has no authority of its own, but only the authority that the first beast—most likely Rome—exercises. Finally, it calls all of those sojourning migrants (οἵ κατοικοῦντες) on the land of Asia Minor to worship the first beast, Rome.

Faced with this host culture's norm of imperial worship, John cannot accommodate it and will not let his community do so either. So from exile,

---

4. Unless otherwise noted, biblical translations are my own.

he writes back to his assemblies insisting against such accommodation on their part and calling accommodators by the names of polemicized heretics of the past: Balaam and Jezebel (Rev 2:14, 20).

John issues a call to *marronnage*, thus to alienation rather than accommodation. All accommodationist migrants reject some of the norms of their host culture. However, accommodationists find ways to negotiate staying within the culture. Alienated migrants pull away from the host community entirely, as John of Patmos advocates. Echoing the call of Jeremiah, John counsels his assemblies: "Come out of her, my people, so that you might not participate in her sins, and so that, from her blows, you may not receive a share, for her sins were joined together until near the sky, and God remembered her wrongs" (Rev 18:4–5; cf. Jer 51:45).

At the same time, John's response is alienation, not marginalization. Revelation affirms John's own culture, both in terms of biblical religion and ethics. The book "samples and remixes"—in the language of hip-hop—the apocalyptic literature of formative Judaism, including Jeremiah, Ezekiel, Daniel, and Zechariah. John of Patmos, like his Rastafarian descendants, turns toward his home culture and uses its notes to chant down Babylon.

If Rastafari's intentional insularity is most like John of Patmos, Miss Lou's accommodationist stance echoes that of Paul of Tarsus. Consider the situation in the Roman colonies of Galatia. As gentiles began to join Christian gatherings, they wrestled with the matter of circumcision, a requirement that had created a class of "God-fearers" (σεβόμενοι) connected to the diaspora Jewish synagogues. In response, Paul, the migrant Pharisee and founder of the Galatian assemblies, welcomes Galatian men into full membership in the community and fictive family *without* having to adhere to his migrant custom. Thus, Paul the Pharisee asserts, "In Jesus Christ, neither circumcision nor the presence of the foreskin are of any power, but faith that is at work through love" (Gal 5:6).

Had Paul simply stood against circumcision and other identifying cultural markers of his own people, one might call him an assimilationist, one turning his back on his own traditions and "passing" for gentile. However, Paul anchors his arguments against circumcision within the same scriptural and cultural tradition that he seeks to nullify. Paul both supports his cultural heritage through the writings of his letters and uses them as the basis for the full inclusion of the gentile. Indeed, Paul invokes none other than the Abrahamic tradition of Genesis, the blessing of the nations (ἔθην) (Gen 12:3, 22:18). In a well-rehearsed *jeu de mot*, Paul connects these "nations" with the Galatian "gentiles," thus declaring his foreskin-bearing hosts a part of the Abrahamic covenant from its inception. In the process, he creates an early Church *patois*, a language of address and theology born from the amalgam of both Jew and gentile, epitomized in his opening words to the churches: "grace *and* peace."[5] These two examples represent some of the stances that islandedness with its constant presence of migration can cause, and perhaps did cause even in the New Testament world. For although Paul is writing to physically "landed" people, his intent is to create "islands" of accommodation while he inhabits, even thrives, in the marine-like borderlands between those ecclesiastic islands, a seafarer until his last voyage to Rome. By contrast, John, who writes from an island and depends on the sea to transport his letters to the seven churches, dreams of a day when the permeability of the sea's boundary, with its threat of cultural annihilation through imperial migration—that is, invasion—ceases to exist, on that day when there is a new heaven, a new earth, *and no more sea*. And, as both Paul's letters and John's Apocalypse are used by communities of faith as Scriptures, their writings, and the others in this and other canons help actual and metaphoric islanders script new strategies of engaging their own diaspora spaces.

There are many other such examples from within the New Testament writings of the importance of the diaspora space of islandedness, whether one means that physically or metaphorically. One might read 1 John as preserving antiemigration polemics, denoting those who leave the "island," namely, "anti-Christs." Certainly, the gospel narratives of the crucifixion suggest a permeability of borders when the cry from Golgotha causes the tearing of the curtain of the temple, death's waves breaching even the sea wall of the "island" Jerusalem. Jesus himself becomes a seafarer, not only because he walks on water, but because, like Paul, he voyages in those permeable boundaries between cities, that metaphorical sea called "wilderness" and "tombs." As the metaphor of islandedness takes hold, its possibilities abound, shaped by the movements of the writers and the original audiences as well as those of the readers in that diaspora space called biblical interpretation.

---

5. Paul's standard greeting, "grace and peace" plays on the typical Hellenistic greeting *chairein* (χαίρειν) and the typical Judaic greeting *shalom* (שלום), which in Greek is *eirene* (εἰρήνη). By combining those two, Paul is modeling his "neither Jew nor Greek" and "both Jew and gentile" ethics.

Islandedness and Further Study: Some Questions

In light of the hermeneutical potential of this perspective on biblical studies, I close by raising several questions, questions that I hope will foster and further discussion. For, these first forays into questions surrounding islands, islanders, and the Bible are themselves diaspora space, and if we would determine the kinds of contours this discussion must take, some interrogation should not only be permitted, but expected.

Much of this essay focuses on islandedness as a state of being that can be engaged through written and/or spoken rhetoric. If this is so, can we imagine a historical islanded theology developing from those on actual islands (like Patmos) or imagined islands (like Qumran)? How might that be different from or overlap with the study of asceticism? Might there be historical or literary critical data to discover from biblical texts and first-century texts that are read from and/or written on islands?

Second, what are the implications of various strategies of islanded cultural formation in diaspora space for those that Elisabeth Schüssler-Fiorenza might describe as at the bottom of the kyriarchal pyramid, those facing multiple—thus multiplied—structures of kyriarchy: gender, sexual orientation, class, age, ability, migrant status, et cetera? How does insularity figure either as oppressive or liberative when "islandedness" reinscribes or concretizes certain forms of kyriarchy?

Two even more pressing questions, questions that hearken back to the initial definition, require consideration. If in fact one island can produce disparate cultural strategies with regard to world, home culture, host culture, and migration, is there such a thing as ontological islandedness? If there is, what might it be? And if not, how then does this discourse contribute to the larger questions of biblical hermeneutics? Is its benefit merely in the metaphor, or is there something more to island discourses?

Finally, this anthology dedicated to islands, islanders, and the Bible must take seriously that there is a significant disconnect between the kinds of work we represented here and some other academic definitions of "island studies." Hay (2006, 30) writes in the first issue of *Island Studies Journal*, "The metaphoric deployment of 'island' is, in fact, so enduring, all-pervading and commonplace that a case could reasonably be made for it as the central metaphor within western discourse." However, these metaphoric descriptions strike Hay as not the subject of islander studies at all. He writes, "I do not believe that they fit within the purview of nissological investigation, which should, rather, concern itself with the *reality* of islands and how it is for islands and islanders in the times that are here and that are emerging" (30).

The question that we must face, then, is what I have done in this essay, and what we have done in this book, "islander studies" at all? If not, what is? And if we claim it to be so, then what is the nature of our *marronnage* from the newly forming discourse around islander/island studies or nissiology? What new questions, thoughts and methods must emerge out of this diaspora space if we are to claim this emerging set of hermeneutical questions as important both to island studies and to biblical studies?

Works Cited

Baldacchino, Godfrey. 2004. "The Coming of Age of Island Studies." *Tijdschrift voor Economische en Sociale Geografie* 95:272–83.

Bennett-Coverly, Louise. 1974. "Back to Africa." Page 194 in *Caribbean Rhythms: The Emerging English Literature of the West Indies*. Edited by James T. Livingston. New York: Washington Square Press.

Berry, John. 2001. "A Psychology of Immigration." *Journal of Social Issues* 57:615–31.

Boomer, Arie and Alistair Bright. 2007. "Island Archaeology: In Search of a New Horizon." *Island Studies Journal* 2:3–26.

Brah, Avtar. 1996. *Cartographies of Diaspora: Contesting Identities*. London: Routledge.

Gillis, John R. 2004. *Islands of the Mind: How the Human Imagination Created the Atlantic World*. New York: Palgrave McMillan.

Hay, Pete. 2006. A Phenomenology of Islands. *Island Studies Journal* 1:19–42.

Marley, Robert Nesta. 1999. "Rastaman Chant." With the Flipmode Squad. *Chant Down Babylon*. Island Records.

Wimbush, Vincent L. 2004. "Introduction: Reading Darkness, Reading Scriptures." Pages 3–43 in *African Americans and the Bible: Sacred Texts and Social Textures*. New York: Continuum.

Building on Sand:
Shifting Readings of Genesis 38 and Daniel 8

Steed Vernyl Davidson

Islands are the ultimate exotic locations. They conjure up images of the getaway escapism that forms the dreams of many metropolitan dwellers. Listening to these fantasies, one would never know that islands are occupied by real persons who lead regular lives. The popular and now long running television program *Survivor* contributes to this mystique. In the show, a group of persons from North America spend time on an island and take turns voting cast members off the island assuming a possession of the island only rivaled by colonial powers. This entertainment-style colonialism stands in front of and alongside of the geopolitical conceptions of islands in the popular imagination. That islands are remote, deserted, always tropical, uninhabited, habitable but only for certain seasons, possessing magical and mystical qualities and occasionally housing noble savages forms the core of Sir Tomas Moore's *Utopia*, William Shakespeare's *The Tempest*, and William Defoe's *Robinson Crusoe*, among other pieces of Euro-American literature in print and other forms.

In the nexus of global power, island communities find themselves overdetermined in ways that define island space sometimes in terms of gain but more often than not, in terms of lack. It should be no surprise therefore that islanders script island space in opposite ways.[1] The narrative of the island creates an alternative consciousness, characterized by continuities, new creations, resistance, revisions, and the like. Obviously, an island ontology exists despite the dominant move of figuring islands

---

1. Works such as V. S. Naipaul's *A House for Mr. Biswas* (first published in 1961) and George Lamming's *In the Castle of My Skin* (first published in 1970) use other structures to stand in for island nations as the means of asserting self-determination and independence.

as both remote and unknown.[2] Spatial dimensions count for something and lend not simply habitable space but identity to those who inhabit that space. The relationship of land space and water space being so proximate on an island shapes a unique outlook as Dara Goldman (2008, 28) sees it: "The island is literally the space of the people and their place in the world is shaped by the relationship between the land and the surrounding waters." Obviously, islands come in various sizes, therefore proximity to major waters would vary. However, a consciousness of being completely surrounded by water exists in island communities unlike that of coastal communities on the continents.

Sand serves as one of the essential features of most island ontologies. The aquatic context of sand and sea defines islands in the tourist brochures, and they may not be too far wrong. The waters of the sea separate and mark out islands as distinct landmasses from any other land space. Kortright Davis (1990, 12) offers that islands, except in few cases, do not possess towering landscapes marked by mountains and valleys. Instead of landscapes, Davis offers that islands are marked by sandscapes. He thinks that sand, as a shifting and changeable constituent, symbolizes the variable circumstances of islands like those in the Caribbean. I find Davis's conception useful in applying sand as the basis for theorizing about island biblical hermeneutics. Davis, however, sees sand as unstable and a marker of catastrophe.[3] I, on the other hand, believe that sand contains more than the angst that Davis sees. Sand provides a place to theorize about island consciousness since it serves not only as the place where the water meets the land, but it is where the outside encounters the inside. Sand is the staging area for island identity as it emerges in relation to other identities. Its variability notwithstanding, sand provides a suitable ground for articulating a hermeneutical lens unique to island communities.

---

2. Derek Walcott (2005, 51) notes that the literary trope of "the unknown island" is an incongruous notion: "Of course, the idea of an 'unknown island' is a contradiction, because either it is unknown or it is an island, meaning a bulk with an at least approximated shape, something coherent and separate."

3. In his chapter titled "The Caribbean Sandscape," Davis (1990) details three crises, one "natural," one economic, one political, that demonstrate the provisional nature of Caribbean society. He concludes the chapter: "The Caribbean sandscape is dominated by the permanence of flux, shifting circumstances and conditionalities, changing fortunes and prospects, realignments and false starts, fresh beginnings and short-lived hopes" (28).

Admittedly, I must face one objection to the use of sand in this work. Surely, some would argue, that sand conjures up all of the exoticizing of island communities that an essay of this nature would reject. However, ontologically the sand signified by the tourist brochure is not the same sand that constitutes the identity of the island. In fact, in most islands that depend upon tourism, a clear demarcation exists between the beaches that tourists frequent and other beaches, as well as clearly marked tourist enclaves. So the sand of this essay need not be confused with the tourist brochure. Of course, the sand of this essay engages the production of those other locations of sand. I am using sand here as the native product of the island, the name of the island, and its existence. Edward Kamau Brathwaite's poem "Islands" captures the constitutive role that sand plays in an island's ontology.

Islands

islands
stone stripped from stone
pebbles
empty shells
chapels of broken windows
no one calls here on the Sunday sand. (Brathwaite 1973, 212)

Amidst the ruins of the islands the sand remains. The sand functions as the site of discovery and meeting. In Brathwaite's thinking it even holds sacred properties.

In this essay, I explore the metaphor of sand as the basis for an island biblical hermeneutic. In doing so, I identify four main character traits that can be drawn from this reflection that can map the outlines of what, for now, I will call island biblical hermeneutics. While ultimately, this essay should have implications for any island community, in its present form its concerns remain restricted to the Caribbean islands. I draw on the thinking and reflections upon Caribbean literature and how they probe the contours of writing from the perspective of islands. I co-opt them here to serve a different purpose of guiding and suggesting some of the main areas of what it would mean to read from island space.

Revisionary Waves

Davis's notion of the shifting nature of sand as a characteristic of island space provides the first marker of an island hermeneutic. Rather than seeing shiftiness as a liability or merely a reactionary response, I look to its virtues and regard this quality of sand as being revisionary. As sand washes on the shores, it creates new shapes, lines, sizes, and textures out of the old shoreline. Sand contains a revisionary quality to both change itself and the things with which it comes into contact. This revisionary quality turns old products into new ones, where at times the relationship between the old and new does not always appear evident. In other words, island space possesses the tendency to creolize not merely in the hybridized fashion of Homi Bhabha.[4] Island space, because of its peculiarities, supports revisions leading towards new creations or what Derek Walcott (1974, 5) regards as a "belief in a second Adam." By this turn Walcott presses for a recreation of the prevailing order without invoking the noble savage motif. With this nuance, Walcott demonstrates that he understands the principalities that are at work in the equation. The essential character of these revisions lie not so much in framing resistance to dominant power, but more so in demonstrating that within island space dominant power remains blunted.

Dominant power stands blunted in island space as the unparalleled force of the sea cannot wash away the sand from the shores of the island. Simon Gikandi (1992, 2–3) agrees with Tzvetan Todorov's assessment of the modernist period, "we are all descendants of Columbus." He offers that in many ways the Columbian "discovery" "invents" the Caribbean. Yet he thinks that this reality, while it shapes the Caribbean, ultimately does not define it. He draws upon Wilson Harris and Derek Walcott who both insist that realities become revised in the Caribbean. Harris posits revision as the response to European realities and a constant process that is "re-visionary and innovative" (Gikandi 1992, 4). Similarly, Gikandi points

---

4. Homi Bhabha (1994, 108, 120) views hybridity as an intentional strategy deployed by colonized people of mimicking colonial forms as a means of indicating its instability. However, he also thinks that hybridity is not a stable condition but one merely used to resist colonial power. Bill Ashcroft, Gareth Griffiths, and Helen Tiffin (1998) note the objections by Robert Young to Bhabha's use of hybridity in a way that invokes racist assumptions. Young argues that hybridity served as a deliberate policy of creating heterogeneous populations in order to weaken "native" stock of people and to prevent them from returning to "primitivism" (120). Rey Chow (1994, 131) similarly raises objections to Bhabha's ideas, regarding them as dominant culture's production of itself. The conception of revision here draws much closer to the thinking of Stuart Hall (1995, 193), who views hybridity as the inevitable mixing of cultures that is "never completed" and the combination of "different cultural repertoires to form 'new' cultures which are related to but which are not exactly like any of the originals."

out that Walcott believes the Caribbean to be inherently inventive pointing to the names, inflections, jokes, folksongs, fables, and so on that mark Caribbean creativity of new things. He sees the claiming and transformation of territory by enslaved people under restrictive conditions as evidence of the tendency to respond with newness to oppressive power (13). Gikandi views Harris's and Walcott's observations of the inventiveness of the Caribbean character as creolization, a response that develops "modernist ways of seeing, knowing, and representing their dislocated culture" (14). Although resistance and survival may feature in this revisionary trait, its main preoccupation is thriving in a new environment and producing the new things required for that changed circumstance.

Like the sand washed to the shores of the islands, the Bible comes as a strange product. Like the wave of "discoverers" that come to the islands, the Bible enters with the presumptions of dominant power. And like the various transplanted and displaced people who take root in the islands, the Bible requires adjustments and revisions. Evidently, this means contextualization of the Bible and relevant reading practices. But contextualization does not get to the inventive and creolizing tendencies of island space. Contextualization assumes the retention of original forms, original concepts, and evidently original power assumptions and original hierarchies.[5] Contextualization makes the Bible an expatriate whose real roots belong outside of island space. The Bible in island space no longer has roots anywhere else. It would become what Antonio Benítez-Rojo (1992, 21) regards as a "syncretic artifact" that he describes as "a signifier made of differences."[6] Consequently, island hermeneutics not only rereads the Bible but also re-creates it within the confines of island

---

5. The following instance related by Edward Kamau Brathwaite amplifies what I mean by contextualization here. Brathwaite (1983, 274–78) explores the creation of Caribbean literature using what he calls "nation language." He laments the fact that some poets in order to be "universal" abandon the nation language. Using the example of Claude McKay's poem "St. Isaac's Church," he shows how the poem in "form and the content are very closely connected to European models." In fact, Brathwaite offers that the connections are so close that only the poet's voice reading the poem makes it authentically Caribbean.

6. Benítez-Rojo (1992, 21) explains syncretic artifacts as the product of melting pot communities, but he alters the conception of how this normally works: "Syncretic processes realize themselves through an economy in whose modality of exchange the signifier of *there*—of the Other—is consumed ('read') according to local codes that are already in existence; that is, codes from *here*. Therefore we can

space. Island hermeneutics goes beyond paying attention to displaced people in a foreign land like Joseph, Esther, or Daniel and finding kindred experiences with them. Rather, it makes Saul, David, and Solomon into islanders. It revises Paul, Jesus, and the disciples within the confines of island space by creolizing them. I am talking about more than simply hyphenating their identities or foregrounding their crosscultural skills. Island hermeneutics involves reading and producing readings of texts that engages them as outsiders while asking them to fit in, to thrive, and to take root in the new environment. The Bible that lands on the sands of the Caribbean islands becomes something new as the Anancy stories of West Africa became new tales, as the village griots became calypsonians, as the masquerade of France became carnival, as Bhangra of India became chutney.[7]

Replays

The second trait of sand that aids in island hermeneutic is its doubled quality. Sandscapes are inherently open spaces, but at the same time bounded spaces. Sandscapes mark open spaces that make arrival easy, but also facilitate departure. Sand also marks the edge of the beginning of the land of which there is little or the end of the sea of which a great deal exists. And still that relationship of land and sea can be overturned depending upon perspective. Harris (1981) thinks of the Caribbean as a gateway through which dislocated people pass. He conceives, with the help of Brathwaite, of the limbo as more than a dance or tourist attraction, but as a signifier of the journey to the island. Brathwaite equates the limbo with the cramped experiences of the Middle Passage from which enslaved Africans emerge to the relative freedom of island space. He sees the dance as a reenactment of that journey. Harris takes this further and associates limbo with the Anancy spider and the necessity of metamorphosis that all arrivants to the

---

agree on the well-known phrase that China did not become Buddhist but rather Buddhism became Chinese" (emphasis in the original).

7. Brathwaite (1983) describes the defining role the pentameter has played in English poetry from as early as the time of Chaucer and the attempts by various poets to break out of its hold using mostly linguistic innovations. He notes that the pentameter assumes a particular experience, and it is not the experience of a hurricane. Consequently, he offers that the nation language of the Caribbean "largely ignores the pentameter" (265).

islands undergo. Harris speaks of the limbo as the "gateway or threshold to a new world" (26). He sees it as "native" to the Caribbean, as the essential marker of transcending "great peril" though with the "strangest capacity for renewal" (27). Harris thinks of the rehearsal of the displacement evident in the dance as the critical aspect of the limbo that marks the essence of island space. Not only the emergence into openness gets staged, but also the confinement and restriction. Limbo juxtaposes the old and new, restriction and freedom, oppression and self-determination. Just as the sand marks spaces of arrival, they serve also as spaces of departure. These sandscapes in reality are sites of openness but also sites of closure, evidence of insularity but also markers of vulnerability, openings for rebirth but also points of no return. Harris offers that limbo "re-play[s] a dismemberment of tribes ... and a curious psychic re-assembly of the parts of the dead muse and god" at the same time (28). Island space admits to a doubleness that performs several functions at the same time. Gikandi (1992) observes that Caribbean writing engages both dislocation and imaginary correctives as a means of subverting imperialism. Island hermeneutics participates in this doubleness.

This doubleness that lends itself to island hermeneutics from the openness of sand calls attention to seemingly contradictory elements that hold together in a single reality. Island hermeneutics refuse to engage in binary thinking or the mutual exclusiveness that Benítez-Rojo (1992, 11) calls "the diachronic repetition on an ancient polemic."[8] Louise Bennett's poem "Dutty Tough" characterizes this doubleness.

> Sun a shine but tings no bright;
> Doah pot a bwile, bickle no nuff;
> River flood but water scarce, yawl
> Rain a fall but dutty tough.[9]

Through a list of contradictions, Bennett describes a society where the markers of prosperity may be foregrounded but the opposite serves as

---

8. Benítez-Rojo (1992, 11) here refers to the reduction in philosophical thought to a choice between Plato and Aristotle. He regards this as an oversimplification that ignores the contribution of other Greek philosophical constructions that results in acceptance of this binary as "the limit of the tolerable."

9. Bennett lists here signs of prosperity that stand alongside evidences of want and lack: brilliant sunshine but still darkness; a boiling pot but insufficient food; rivers in flood but scarce water supply; abundant rains but parched earth.

the more controlling reality. Yet, nonetheless, these two realities converge and the closing line of the refrain, "rain a fall but dutty tough," exemplifies not simply ambivalence or ambiguity, as Bhabha would have it,[10] but doubleness. Doubleness here contains these opposites that both critique and undermine the essential power of oppression. In Bennett's poem, both presence and absence critique the general lack of resources or better put the maldistribution of resources. In the same way, the limbo reenacts the Middle Passage and the arrival off of the ship as subversions of human trafficking that supports the plantation economy.

In island biblical hermeneutics, doubleness characterizes the reading process. Beyond the identification of polar opposites in texts, island biblical hermeneutics probes how doubleness confronts various issues. For instance, reading Gen 38, island hermeneutics pays attention to the death and birth of sons, noting that they all have the same father rendering the one woman wife and mother of brothers. Rather than merely taking comfort in the new birth as the ultimate resolution of an oppressive system, island hermeneutics' regard of doubleness calls attention to the system of levirate marriage, the family relations that spring from these marriages, and the place of women in these (re)constructed families. Since attention is drawn to both the death and birth of sons, island hermeneutics focuses attention on Judah, the site where both of these poles converge. Judah's exercise of his paternal rights with regards to his dead sons portends what his behavior would be like with regards to the new sons born to Tamar. His declaration of his failure in Gen 38:26 relates only to his treatment of Tamar with respect to Shelah; it offers no indication of his response to Tamar as mother of his other living sons, Perez and Zerah. That the mother, Tamar, will fare any differently than the wife Tamar under the exercise of patriarchal power by the father-in-law/father Judah serves as a site of inquiry for island hermeneutics. In doing so, island hermeneutics uses both the death and birth of sons to interrogate the system of levirate marriage in the same way the limbo reenacts both the Middle Passage and the landing in the island to subvert the slave trade.

---

10. Bhabha (1994, 132) articulates that ambiguity functions in colonial society as a means of dissimulating and defying categorization. He sees it as a deliberate strategy of obfuscation designed to resist colonial power.

The Motion of the Ocean

The third feature of sand that helps theorize island biblical hermeneutics lies in sand's proximity to the sea. Sand exists as the product of tidal movements. The constant back and forth of the tides, and the movement of the waves, helps to produce sand. Sand serves as one of the testimonies of tidal movements. This tidal quality shapes the rhythm, movement, and measurement of time in island space. Benítez-Rojo (1992, 11), in ways similar to Davis, defines Caribbean culture as more attuned to the sea than to the land: "not terrestrial but aquatic." Benítez-Rojo goes on to describe the impact this tidal orientation has upon conceptions of time in the Caribbean: "a sinuous culture where time unfolds irregularly and resists being captured by the cycles of clock and calendar" (11). Here Benítez-Rojo thinks of more than simply the distinction between linear and circular time. His notion of circularity lies more in his conception of the Caribbean as an island that repeats itself. Drawing upon modifications of chaos theory,[11] Benítez-Rojo offers an application of Caribbean reality that resembles interlocking systems engaging in flows and interruptions. The rhythmic interaction of flows and interruptions creates feedback where every repetition differs from the previous one. Benítez-Rojo sees the Caribbean as an island that repeats itself, where the rhythmic pulses of repetition, reproduction, growth, decay, ebb and flow, and the like mark time and form the basis of history. In Benítez-Rojo's conception, history is not a series of moments, as Walcott laments in his essay *Isla Incognita*. Walcott (2005, 57) describes the historical moments of European invasions that form the highlights of the history of conquered people. He wittingly speaks of Lewis and Clarke beholding the west coast of North America and asks what are they beholding and then answers, "They behold the images of themselves beholding." In a similar vein, Walcott (1974, 4) suggests, "The slave surrendered to amnesia ... is the true history of the New World."

The departure from linear history for both Walcott and Benítez-Rojo does not serve as the automatic escape hatch from European overdeterminism in reality and historical recording. Amnesia provides the route for that rejection for Walcott.[12] However, they both conceive of time and history

---

11. Benítez-Rojo (1992, 3) does not admit to using chaos theory only a "new scientific perspective" and not that which is "conventionally defined."

12. Walcott matures his notion of amnesia as a response to history over time and articulates that he is not advocating escape from a history that cannot be faced but

outside of linear categories and events but within the rhythms of experiences. Walcott (2005, 54) notes that northern climates are more seasonally rhythmic than the monoclimate of tropical islands. In his estimation that should mean that islanders have one "common temperament." Implicit in Walcott's analysis lies the conclusion that those seasonal changes do not produce cultural temperaments that are attuned with the movements of nature, the variable interactions of forces that can alter the teleological triumphalism that characterizes European history. Walcott goes to island space to theorize about history, as it may exist, in his poem "The Sea is History." As a retort to the claim of the Caribbean being ahistorical,[13] Walcott narrates how the sea shapes Caribbean history and with his invocation of several biblical books, themes, and events, we can add, salvation history.

> Where are your monuments, your battles, martyrs?
> Where is your tribal memory? Sirs,
> in that gray vault. The sea. The sea
> has locked them up. The sea is History.

Ironically, Walcott invokes several events of colonial history but overwrites these with biblical data to create the alternative history. The sea remains his true source for history.

> First, there was the heaving oil,
> heavy as chaos;
> then, like a light at the end of a tunnel,

---

that the notion of history is inappropriate to the Caribbean. Edward Baugh (2006, 10) records Walcott: "In the Caribbean history is irrelevant, not because it is not being created, or because it was sordid; but because it has never mattered. What has mattered is the loss of history, the amnesia of the races, what has become necessary is imagination, imagination as necessity, as invention."

13. Baugh (2006, 8) notes Walcott's struggle with the nineteenth-century British historian, James Anthony Froude's assessment that since the Caribbean can show no achievement then it really has no history. Although Walcott uses the term history, it remains unclear whether he think it exists in relation to the Caribbean. In the poem he writes, "Then came the white sisters clapping / to the waves' progress, / and that was Emancipation—/ jubilation, O jubilation—/ vanishing swiftly / as the sea's lace dries in the sun, / but that was not History, / that was only faith, / and then each rock broke into its own nation." John Theime (1999, 161) reads Walcott as constructing a "tribal memory," and the elements of that memory serve as evidence of faith. Baugh (2006, 9) writes of Walcott "going beyond history."

the lantern of a caravel,
and that was Genesis.
Then there were the packed cries,
the shit, the moaning:

Exodus.

The poem ends with the assertion that in paying close attention to the island environment one can see history "really beginning."

The historical orientation of island space concerns itself not so much with trajectory, linear versus cyclical, but rather with rhythm. In fact, Benítez-Rojo (1992, 24) views the repeating island as transhistorical in the way Walcott sees it as ahistorical. For both of them the normal categories of history do not apply. Time in island space responds to tidal movements that are consistent and regular but unpredictable and unrepeatable. Time in island space, therefore, stands as an amalgam of events, experiences, encounters, incidents, understandings, insights, and the like that go together to shape history; not as a narrative with a predictable plot and final dénouement but of events related in some ways to one another but different in outcome and scope. As tides come in pulsing waves each with a different size, shape, and intensity and over time produce sand, island space marks time and creates history in tune with the rhythm of its environment.

An island biblical hermeneutics operating on time marked by the sea finds the Bible a strange place with its foreboding of the sea and its imagining that the sea be no more (Rev 21:1). Tidal time exists even with distance from the sea since it is the orientation to island space that remains determinative. Therefore, biblical texts that contain no sense of the sea, apart from a few narratives in the Gospels, can be read from island space. Reading Gen 38 in tidal time reveals the absence of definitive time markers that will lead to the construction of a history. Time is marked more in terms of events, nondescript periods, with actions melding into actions that blur the temporal distinctions. The narrative follows a genealogical development that moves from Judah in Gen 38:1 to concerns about his third generation by Gen 38:8. The story begins with Judah's relocation dated as "at that time." This easily overlooked notice calls attention to the motivation for Judah's move at that time. Is "that time" referring to Judah's heroic "rescue" of his brother Joseph from his bloodthirsty brothers and his brokering of the sale to the Ishmaelites (Gen 37:26)? Or does it refer to his inconsolable father mourning the loss of a son, unknowing that his other sons contrived to sell him into slavery? Is moving at "that time" propitious for Judah or not? How is he acting in terms of tidal time? Interestingly, we are told that he "went down from his brothers" (Gen 38:1). Has the tide receded in a way favorable for Judah or has it come in? Likewise, the next time that Judah goes on the move, the text reports, he "went up to Timnah" (Gen 38:12). There he encounters Tamar, who, in his estimation, was a sex worker. Are these two movements by Judah in response to favorable tides or will the story prove him to be pathetic in terms of tidal time?

The narrative of Gen 38 attempts to be ahistorical for the most part and stands outside of recorded time. It purports to be uninterested in recorded time. Simple notices of conception, birth, and naming mark the rapid succession of Shua's births. This formula is repeated in Gen 38:3, 4, and 5 for each son's birth. But as Benítez-Rojo avers, repetitions produce new iterations; we observe the newness in each child's birth announcement. Judah names Er, while Shua names Onan and Shelah. The presence of "again" may lull the historical mind into sameness, but in tidal time, "again" serves as an alert for something momentous. So when we have "yet again" in Gen 38:5, we receive even more momentous news. Shua is in Chezib when she gives birth. This serves as the first time that the environment enters into the story. Time now unfolds in a particular place, and that place exerts force upon the story. The place where Judah relocates at the start of the story remains unknown, and consequently his actions there stand as relatively inconsequential in the story. However, his second movement in the story to a named place produces consequences in the story. Time in island space emerges in response to the environment, the sea, but here, place. Time unfolds against the movement of the tides of the sea but in this story against the background of contact with the lived space.

While most of the narrative stands outside of historical time, one notice abruptly brings the focus of the story into historical time. Unlike the nondescript markers of "in the course of time," "Judah's time for mourning" (Gen 38:12), or "when the time of her delivery came" (38:27), we find the stark notice "about three months later" (38:24). This notice introduces legalism into the narrative that condemns Tamar for prostitution. That the claim of transgressing the community mores comes alongside of the marking of time by a set of socially constructed norms looks curious from the standpoint of tidal time. That the assertion of this claim ultimately fails as Tamar produces evidence she acquires at Enaim, a named place where tidal time unfolds in relation to place suggests the ineffective of the calendar, the legal prescriptions, and the attempt to capture perceived female sexual misconduct as historical fact. The resumption of ahistorical markers in Gen 38:27 in relation to the birth of the twins also underscores this failure.

Shorelines

In locating a fourth trait for theorizing about island biblical hermeneutics, I return to the shifting nature of sand that Davis invokes. As before, I deal with this as a virtue and observe that the shifting sands also serve as the boundary, marking the promise of possibility. Sand may not make good foundations for building, but it provides good support in building construction. In this way sand points to endless possibilities. Sand marks the boundaries between land and sea, firm ground, and water and therefore indicates the start of possibilities. As Goldman (2008, 10) observes, sand shapes the form, definition, and identity of the island. Islands contain clear boundaries formed by their natural geographies, and while the continental predisposition lies in looking out towards the sea for horizons that chart possibilities, island space sees possibilities as beginning at the shore. Within the confines of limited space, limited resources, and at times separation, island communities learn to thrive and establish themselves. Goldman offers that "the island is literally the space of the people" (28). The piece of earth surrounded by sand reveals possibilities from the perspectives of insularity, emancipation, and the antiapocalyptic.

In many instances, insularity stands as a trait in island space that should be eliminated. At times this trait emerges as a problematic characteristic of islanders. Goldman (2008, 7) accounts for insularity as a product of island space. She notes, though, that with vulnerability comes insularity as another product of geographic space. Nonetheless, she argues that island space produces a clear subjectivity that asserts itself as self-determination. Clear national identities underlie insularity given that in most cases island and nation are coeval. Goldman explores how insularity works at the level of the Greater Antilles.[14] She shares that insularity surfaces as a pervasive trope in the literature of these islands. Since she balances insularity with vulnerability, she remains careful not to over articulate this as a sacred

---

14. This would include the islands of Cuba, Jamaica, Hispaniola (Haiti and Santo Domingo), and Puerto Rico. Her concerns, though, lie with the Spanish speaking Antilles—Cuba, Santo Domingo, and Puerto Rico.

virtue. Therefore, she notes the limitations and the shifting ground out of which insularity emerges (48). At the same time, she calls attention to insularity as the productive center of community, identity, and engagement with the world. Island space engages with possibilities not in an idealistic utopianism but within the confines of vulnerabilities and limitations.

Emancipation serves as another perspective to view possibilities in island space. Davis opts for this term as unique to the Caribbean historical realities. In doing so he moves away from the more popular term "liberation."[15] These two terms, liberation and emancipation, envisage a world of possibilities where inequities get sorted out and the prevailing structures of power are not simply inverted but dismantled. Patrick Taylor positions Caribbean ontology as pressing for this goal. In distinguishing myth from narrative, he offers narrative liberation as a Caribbean possibility. Taylor views myth as imaginative inversions of power, like those in the tales of the Anancy spider that triumphs over the more powerful animals. Anancy's victories, though, last only for a time, since the structures of power that define him as weak remain, or if the victories last longer they only replace one dominant power at the top of the chain with another. On the other hand, Taylor (1989, xii) sees liberative narrative as offering a radical departure from current realities: "The narrative of liberation reveals the limits of the struggle for a hallowed ancient past, the endurance of a wretched present, or the leap toward a utopian future; it engages the process of historical transformation with a view to the possibility of creating a society based on human mutuality."

A firm grasp of history lies at the heart of Taylor's conception. This prevents his thinking from devolving into naiveté. At the same time, it involves possibilities within the realm of reality. In pointing out that his views are not merely abstract, Taylor insists that the daring vision of mutuality that he calls for prevents liberative narratives from being subject to "mythical closure" (6).

Island space then conceives of possibility in decidedly antiapocalyptic terms. Taylor's liberative narrative works against apocalyptic that holds the simple binary where the oppressed will become empowered through an

---

15. Davis (1990, 6) explains that he prefers the term emancipation, since it serves as a pointer to the historical search by Caribbean people for "ascendancy, control and emancipation." By using this term, he both invokes and disrupts the historical narrative that asserts moments of emancipation or independence and draws attention to prevailing forms of "neocolonialism and inertia."

inversion of power. In Taylor's categories, apocalyptic is subject to mythical closure. The antiapocalyptic perspective also emerges from what Benítez-Rojo sees as the inherent antiapocalyptic character of the Caribbean. Using sexual images, he observes that the Caribbean does participate in the "vertical desires of ejaculation and castration" (Benítez-Rojo 1992, 10). That is to say, Caribbean culture breaks out of the simple binaries of power distributions that produce winners and losers. His retort, "Here I am, fucked but happy" in response to crises may be seen as typical Caribbean nonchalance, but Benítez-Rojo views it as the character of island space. He thinks of Caribbean people as "People of the Sea" and therefore of island space as marked by the sea. This defining quality shapes a rhythmic performativity in island space that Benítez-Rojo believes "sublimate[s] violence" (20). He garners this confidence from the notion that technological and capitalist consuming cultures produce violent responses unlike the sea that explores other possibilities for change and coexistence.

Apocalyptic literature poses a curious challenge to island biblical hermeneutics if this hermeneutical perspective assumes an antiapocalyptic character. In reading Dan 8, the focus remains on the vision in the first part of the chapter while treating the explanation of the vision in the latter portion of the chapter as a secondary text. This vision of ferocious animals immediately enters island space as myth, but only disturbs island space due to its pointlessness. This note of bother emerges in the question of verse 13 that suggests impatience with the hiatus in the sanctuary. The matter appears to be dropped after an answer is given in verse 14 and does not remerge. The "holy one" of the question perhaps comes from the group identified as "all beasts" of verse 4 that experiences powerlessness against the ram. Apart from the notice of their vulnerability in verse 4, this question also demonstrates a level of vulnerability. The question also moves in the direction of insularity, since the site of the sanctuary appears to be only collateral damage in the confrontation between these extraordinary beasts. The movement of the battle to a site in verse 11 contains the note of it being merely incidental in the larger scheme of things. Island biblical hermeneutic expresses concerns for this easily overlooked group known as "all beasts," presumably a large portion of living beings, that needs to assert its identity in view of the strength of these two other creatures. From the perspective of insularity the description of the ram and the goat suggests that they have reached their potential. In the case of the ram, it reaches this potential and then falls. In the case of the goat, the answer in verse 13 indicates that its peak would arrive as well. Possibilities for constructing the group of other beasts exist in the narrative, because they need to define themselves over and against these two other creatures since they are restricted to the narrow space of a single verse. Additionally, the possibility for survival for these other beasts exists in adopting a position of getting out of their way. Both creatures prove themselves self-destructive or act in ways that their own ambitious will destroy them. These are traits that are lacking in the other beasts and the way in which they can assert their subjectivity. Of course, the narrative never indicates the total eradication of these other beasts so they can adopt the, "Here I am, fucked but happy" stance of Benítez-Rojo. Perhaps, the same could be said of the ram since its fate in verse 7 remains unclear.

Reading from the perspective of insularity that Goldman holds together with vulnerability, island biblical hermeneutics notices that both the ram and the goat do not claim total dominance. In the case of the ram, it only changes in three directions (v. 4); it is unable to completely circumnavigate. Similarly, even though the goat sports four main horns that touch the four winds (v. 8), the little horn that grows out of one of the horns only moves in three directions (v. 9). The descriptions of these two animals convey power and strength but lack any conceptions of their vulnerability and therein lies the possibility for their demise; the ram in verse 7 is defeated by the goat, and the goat in verse 12 is destined to a mythical existence of continual prosperity as illusory as the search for El Dorado or the Fountain of Youth.

The close of the vision offers an antiapocalyptic note for island space. This note envisages a return to a normal state in verse 14 without the use of violence. This vision contains no resolution of the problematic goat that seems so fearsome at first. However, in the end its defeat comes not through conjuring a more powerful animal but simply a question of how long it will last. The question sublimates the violence that rages in the earlier portions of the visions and offers new possibilities for those affected by the ram and the goat. The explanation of the vision in verses 15–27, however, moves this text into the apocalyptic column from the perspectives of island biblical hermeneutics. The animals are no longer animals but actual historical empires. Reading verses 1–14 requires little affective involvement from the reader, while the second portion of the narrative asks readers to make choices based upon their associations with these named empires. Further, the explanation invokes the violence missing in verses 13–14 that resolves the crisis. The king of Greece named in verse 21 will be a purveyor of violence (vv. 23–25) and ultimately will be undone without hands. The violence against the king of Greece is conveyed in starkly economic terms of merely three words in Hebrew closing on the *niphal* form of the verb to indicate his complete helplessness. Unlike the promise of restoration to "its rightful state" in verse 14, the end of the explanation offers no vision of the future. It implies the eradication of the problem but offers no means of how the future works to ensure the elimination of these forces. The absence of a look at the systems that produce the ram and the goat appear in both sections of the chapter, subjecting it to what Taylor calls mythical closure. In the end, island biblical hermeneutics finds Dan 8 both engaging and distancing.

Conclusion

Island biblical hermeneutics built on sand pays more attention to how to read the Bible in island space than with what is the Bible. This essential difference marks the nature of hermeneutical engagements in many other spaces. The insistence on reading the Bible not simply as a transient visitor, but as an arrivant with possibility for long term residence opens up the question as to how the Bible functions in island space. Precisely, because of the uniqueness of the social location of the island, island biblical hermeneutics stands as a pressing need. The geographical uniqueness of sand that defines multiple island spaces offers possibilities for articulating this peculiar set of strategies for reading the Bible. The four elements of this hermeneutical strategy, which I described here, require both reader and text to engage island geographies seriously. Revisioning waves questions the identity of the biblical text as much as it does the reader. What difference does island space make when reading the Bible? "Doubling Replays" interrogates vision by adopting the default position that what is seen often times plays tricks upon the mind. "Motion of the Ocean" examines conceptions of time and thereby challenges the historiography that writes the narrative of the island and its inhabitants. Finally, "Shorelines" explores aspects of power by positioning island space as antiapocalyptic by envisioning radically altered practices of power.

For some, island biblical hermeneutics may offer little that appears innovative. The test of innovativeness lies not so much in what new things that island space can produce about the Bible; rather, innovativeness comes from a clear articulation of island space so as to make the Bible new in that space. Close attention to the space that is an island makes this possible. Can anyone move within island space and not get sand in their shoes? Sand serves as one of those critical sites from which to theorize island ontology.

Works Cited

Ashcroft, Bill, Gareth Griffiths, and Helen Tiffin. 1998. *Key Concepts in Post-Colonial Studies*. London: Routledge.
Baugh, Edward. 2006. *Derek Walcott*. Cambridge: Cambridge University Press.
Benítez-Rojo, Antonio. 1992. *The Repeating Island: The Caribbean and the Postmodern Perspective*. Durham, NC: Duke University Press.
Bhabha, Homi K. 1994. *The Location of Culture*. New York: Routledge.
Brathwaite, Edward Kamau. 1973. *The Arrivants*. London: Oxford University Press.
———. 1983. *Roots*. Ann Arbor: University of Michigan Press.
Chow, Rey. 1994. "Where Have All the Natives Gone?" Pages 125–51 in *Displacements: Cultural Identities in Question*. Edited by Angelika Bammer. Bloomington: Indiana University Press.
Davis, Kortright. 1990. *Emancipation Still Comin': Explorations in Caribbean Emancipatory Theology*. Maryknoll, NY: Orbis Books.
Gikandi, Simon. 1992. *Writing in Limbo: Modernism and Caribbean Literature*. Ithaca, NY: Cornell University Press.
Goldman, Dara E. 2008. *Out of Bounds: Islands and Demarcation of Identity in the Hispanic Caribbean*. Lewisburg, PA: Bucknell University Press.
Hall, Stuart. 1995. "New Cultures for Old." Pages 175–214 in *A Place in the World? Places, Cultures and Globalization*. Edited by Doreen Massey and Pat Jess. Oxford: Oxford University Press.
Harris, Wilson. 1981. *Explorations: A Selection of Talks and Articles 1966–1981*. Mundelstrup, DE: Dangaroo.
Lamming, George. 2004. *In the Castle of My Skin*. Ann Arbor: Ann Arbor Paperbacks.
Naipaul, V. S. 2001. *A House for Mr. Biswas*. New York: Vintage.
Taylor, Patrick. 1989. *The Narrative of Liberation: Perspectives on Afro-Caribbean Literature, Popular Culture, and Politics*. Ithaca, NY: Cornell University Press.
Theime, John. 1999. *Derek Walcott*. Manchester: Manchester University Press.

Walcott, Derek. 1974. "The Muse of History." Pages 1–27 in *Is Massa Day Dead? Black Moods in the Caribbean*. Edited by Orde Coombs. Garden City, NY: Anchor.

———. 2005. "Isla Incognita." Pages 51–57 in *Caribbean Literature and the Environment: Between Nature and Culture*. Edited by Elizabeth M. DeLoughrey, Renée K. Gossom, and George B. Handley. Charlottesville: University of Virginia Press.

Island-Marking Texts:
Engaging the Bible in Oceania*

Nāsili Vaka'uta

> Human reality is a human creation: And ... if we fail to create our own, someone else will do it for us. (Hau'ofa 1993, 128–29)

This essay shares some insights on biblical interpretation from the standpoint of an islander, hence the title, "Island-Marking Texts." Island-marking is my translation of the Tongan phrase *lōlenga fakamotu* ("doing things in the ways of the *motu* [island]") or *lau fakamotu* ("reading island-wise, as an islander"). Island-marking texts refers to a reading of the Bible that *arises out of island contexts, shaped by island cultures and values, gives privilege to island knowledge systems (epistemologies) and languages, reads the Bible through island/oceanic lenses, takes account of critical issues that confront islanders, and serves the interests of the islands and islanders.*

This chapter offers a Tongan concept, *fale-'o-kāinga*, as a framework for "island-marking texts" and outlines how *fale-'o-kāinga* makes biblical hermeneutics meaningful and relevant for readers in Oceania. *Fale-'o-kāinga* is local to Oceania, and it encourages relational versus colonialist attitudes toward biblical texts.

Island-Marking

"Island-marking" utters a call, first of all, to negotiate and claim a space for island perspectives and hermeneutics—be they Pacific, Caribbean, Taiwanese, et cetera—and, secondly, to promote more openness in biblical

---

* A draft of this chapter was presented at the meeting of a new Society of Biblical Literature unit, Islands, Islanders, and Bible, at New Orleans, Louisiana (22 November 2009).

studies to our *lōlenga fakamotu*. I am not endorsing a move towards hermeneutical apartheid; such a move would create island versions of colonial scholarship and jeopardize the cause I promote herein. I am simply advocating solidarity amongst island scholars to free ourselves from the epistemological domination that dictates biblical scholarship.

Why island-marking? What permits such a move? In recent years, the terrain of biblical studies has been altered considerably by various paradigm shifts, two of which are postcolonial and ecological hermeneutics. Both paradigms opened up new routes for different ways of reading and duly raised awareness to the global environmental crisis and the impact colonization has had on colonized contexts the world over. In the area of postcolonial biblical criticism, a significant contribution have been made through the works of R. S. Sugirtharajah, Fernando Segovia, Tat-siong Benny Liew, and many more. Norman Habel and Earth Bible cohorts advocated justice for the earth and the earth community when reading the Bible. A number of practitioners from Oceania embraced both reading perspectives with open arms.

On a personal level, I appreciate the contributions made by both camps, but I have some reservations, and two of them are based on the realities that confront islands in Oceania. First, some islands in Oceania are still under foreign rule—for example, Hawaii, Tahiti, New Caledonia, American Samoa, Aotearoa, Guam, Palau, and more. Postcolonialism, promising though it sounds, creates a false reality for the inhabitants of these colonized islands. The reality is that no matter how much they read the Bible from a postcolonial standpoint, it will not negate the fact that colonialism is very much alive at home, in Oceania. Choices of those nations are limited, and to make things worse, some cultures and languages are at the verge of extinction.

Second, the devastating effect of climate change has shaken the region to its core. A tsunami damaged the islands of Upolu in Samoa, Tutu'ila in American Samoa, and Niuatoputapu in Tonga. Many islands like Tuvalu are at the mercy of nature. The sea level rises constantly, and there would be nowhere to hide if tides rushed inland. The issue at stake for the people of these islands is not how to save the earth and nature (as promoted by ecological hermeneutics), but how to survive the earth and the wrath of nature.

Unless there is more inclusion of island perspectives in biblical scholarship, and unless mainstream scholars allow islanders to think for themselves and to read texts as islanders, international gatherings like the Society of Biblical Literature will only foster scholarship that is irrelevant to people of the islands. Cornel West, with reference to what he calls the "new cultural politics of difference," advises that it is about time

> to trash the monolithic and homogeneous in the name of diversity, multiplicity and heterogeneity; to reject the abstract, general and universal in light of the concrete, specific and particular; and to historicize, contextualize and pluralize by highlighting the contingent, provisional, variable, tentative, shifting and changing. (West 1993, 3)

Islanders in Oceania appreciate diversity, because Oceania is one of the most culturally diverse regions on the earth, with hundreds of distinct cultures, languages, and knowledge systems, most of which had existed for thousands of years. We also have diverse belief and value systems that precede the arrival of Christianity on our shores.

Oceania is more than just exotic places, with white sandy beaches, and half-naked hula girls having coconut shells on their chests. There are more to its peoples than being hard-running brown rugby players with natural flair and talents and big tattoos on their bodies. Islanders can think and theorize; we can negotiate and imagine for ourselves, rather than being imagined by others for us. In the case of the Bible, we have at our disposal a rich stock of metaphors, concepts, and values that can be utilized as lenses and methods for analyzing texts. Hence in the next section, I articulate a Tongan concept, *fale-'o-kāinga,* as a framework for "island-marking texts."

Fale-'o-Kāinga

Fale-'o-kāinga is a combination of two words *fale* and *kāinga*. The word *fale* generally refers to a house or a dwelling place, the *oikos*. *Fale*, however, means more than a physical structure that we built. It refers to any group of people with common interests and a shared sense of belonging—groups like families, communities, churches, and so forth. *Fale* in this sense is a commonwealth of people who are united by their responsibility and commitment to each other. *Fale* indicates presence (as opposed to absence) or being at home. From this we have the cognates *fafale* (mutuality and intimacy), *fale-taha* (sharing space), *falenga* (to dwell or to accommodate), and *lotofale'ia* (the warmth of dwelling together).

We also speak of *fale* metaphorically. First, *fale* is an ecological space. Wherever we are in the world, we belong to a household that we share with other creatures. In Oceania, we have "a huge watery house." We do not own that house; we belong to it. That belonging comes with a responsibility that, if not observed, would jeopardize the whole household and the well-being of those with whom we share that space.

Fale is also an economic space that is managed and maintained not by some strange philosophies, but by a cultural network of relation and exchange that we call *kāinga*. *Kāinga* basically refers to kindred, but it also indicates that which is relevant, applicable, and well connected. Within that network, goods and services are exchanged, and that contributes to strengthening the relationship within the *fale*. *Kāinga* is also the major driver behind remittances sent to the islands by islanders residing overseas to fulfill their *kāinga*-related duties. In return, people in the islands sent goods as tokens of appreciation. That exchange amongst *kāinga* transcends borders and crosses boundaries.

At the heart of the *kāinga* network is a core value that is characteristic of most, if not all, cultures in Oceania: *reciprocity* (Tongan, *tauhi vā*; Samoan, *teu le vā*). *Tauhi vā* elevates distribution above consumption, sharing above accumulation, peaceful coexistence above domination, communal well-being above individualistic interests.

To ensure sustainability within the *fale-'o-kāinga*, there are some simple guiding principles to observe and practice, and they require neither expert advice nor a university qualification in order to understand.

The first principle is *faka'apa'apa*. *Faka'apa'apa* requires that each member in the *fale-'o-kāinga* must show *unreserved* respect for each other. When there is no respect, the land, the sea, and the air are products to sell in order to stimulate growth. When there is no respect, we run the risk of prostituting ourselves or pimping others for our own gain. When there is no respect, we measure our fellows not in terms of their human value, but in terms of their employability, efficiency, and productivity. Respecting nature and each other is urgently required in order to sustain our ecological and economic space, our *fale*.

The second principle is *fakapotopoto*. This principle calls for wise management of resources. It requires frugality and thriftiness. The global ecological and economic crises are the result of unwise management and abuse of the scarce resources that we have. We are in a world that is obsessed with the idea that economic growth is the ultimate goal and solution to all our problems. Profits are made by utilizing the cheap labor market of Asia, especially China and India, and their mostly unregulated working conditions. We are in a world that tends to sacrifice common sense on the altar of neoliberalism and the market. Whoever controls the market rules the world in some sense. At the end of the day, poor countries like islands in Oceania continue to be controlled and manipulated economically and politically.

The third principle is *femolimoli'i*. This principle advocates sharing what one has no matter how small it is. I grew up in a household where the income of both parents barely made ends meet, but we managed to survive because my parents made sure that whatever we had was fairly shared. My father (who is no longer with us) was not wealthy, but sharing with others outside our family was very much a part of his life. That is *femolimoli'i*. It is about self-giving, and self-giving is a rare commodity in a world that is dominated by selfishness and greed. We have enough to feed the world, but many people are going to bed hungry every night. To fix that problem, we need self-sacrifice and fair distribution of resources.

The fourth principle is *fua kavenga*. *Fua kavenga* is about fulfilling ones duties and obligations to families, neighbors, and those within the *kāinga* network. Events like funerals and weddings are the responsibility of the *kāinga*. It is about bearing each other's burden and showing that you care. It is about acknowledging that we are *interdependent*. To ignore your duties gives the impression that you do not care and do not need the support of the *kāinga* with whom you share the *fale*. Lack of care for nature and others have produced ecological and economic problems worldwide.

The fifth principle is *mateaki*. In addition to the above, it is important to show that you are committed to *kāinga* and care about the well-being of the whole household. A degree of sincerity and devotion must accompany whatever we do, rather than just acting for acting's sake. *Mateaki* is about sharing with others our entire being without attempting to dominate and dictate their lives. The same attitude is required vis-à-vis nature. We need sincere commitments to nature if there is hope for a sustainable environment. We need sincere devotion to the welfare of others if we are serious about fighting poverty.

The sixth, and final, principle is *loto-tō*. This principle encourages humility. It acknowledges the fact that we do not know everything; we are not better than others; and we are not always right. The opposite of *loto-tō* is *polepole* (pride). *Polepole* is the stuff of egomaniacs and narcissists. *Polepole* is one of the major drivers behind racism, sexism, ethnocentrism, neocolonialism, and neoliberalism. Pride is also the foundation upon which Western attitudes to natives are erected. Lack of humility is evident in the ongoing refusal in some quarters to accept that indigenous worldviews, belief systems, values, and epistemologies can offer valuable insights in areas like academic research, policy-making, and in forums that discuss climate change and economic development. The world will be a better place if the dominant and powerful are humble enough to acknowledge the mistakes they have committed, the problems they have created, and the issues they cannot solve and to accept other views as valid, valuable, and viable alternatives.

Articulating these principles of *fale-'o-kāinga* does not encourage returning to the past or claiming some kind of cultural purity. I am not endorsing a wholesale rejection of existing scholarship. *Fale-'o-kāinga* is articulated simply to promote the fact that, as mentioned above, we have a rich storage of knowledge, values, worldviews, and wisdom that should be brought in to academic discourses and utilize in biblical interpretation.

Rethinking Biblical Interpretation

The attention in this section is twofold: observations and a proposal; both are presented in general terms. Let me begin with the observations, most of which reflect the status of biblical interpretation in Oceania (perhaps the world over). Biblical interpretation, however theorized and practiced, is still an *uneven* and *strange* task for the following reasons:

First, the agendas and frameworks for interpretation are predominantly Eurocentric and colonial, and there is an ongoing preference for that rather than ones based on the treasure that we have—our cultures, our belief and value systems, our texts, our ways of being, and our ways of knowing.

Second, English (or some versions of that language) is still the *lingua franca* of academic conferences and discourses. Even in meetings of people belonging to the same ethnic group, like Tongans, English is still used to cater for the interest of a few English-speaking participants or for a few islanders who are too busy to learn the language of their parents.

Third, biblical scholarship is dictated by the demand of the employment market, and we strive therefore to live up to the expectations of the market rather than what is best and fitting to our diverse cultural contexts.

Fourth, Oceania is excluded in discourses on biblical interpretation as if the region and its people do not exist. The challenge for us in Oceania is not just to decolonize scholarship, but also to "decontinentalize." Island views are drowning under the currents of continental perspectives—from Europe, Africa, Latin America, and Asia.

Fifth, and finally, biblical interpretation is uneven and strange, because we are epistemologically intoxicated and therefore tend to be cautious in what we do to avoid upsetting academia and its traditional norm of scholarship.

To give some balance to the unevenness and strangeness of biblical interpretation, I propose these guidelines derived from *fale-'o-kāinga* as an "island-marking" mode of interpretation:

The notion of *fale* reminds readers to be forever mindful of the fact that we belong to a community, a household. Our reading is shaped by the way we relate to other members of the *fale*, be it human or nonhuman. However we read a text will always have an impact on those with whom we share the *fale* space. It is our responsibility as readers to make sure that our readings will not jeopardize their well-being. It is our responsibility as readers to read with those that are pushed to the fringes by society and texts of the Bible—the poor, the orphan, the widow, and those whose voices are whispering between the lines and lurking at the underside.

The idea of *kāinga* highlights that we (as readers) are ontologically connected and interdependent. As soon as we participate in interpretation, we are entering a network of relations where everyone has a duty and responsibility to each other. This *kāinga* is not defined by any border; it crosses boundaries to salvage those who have been victims of economic exclusions, racial and sexual discrimination, religious intolerance, and scholarly ignorance.

Within the *fale-'o-kāinga*, we are urged to read the Bible with respect (*faka'apa'apa*): for readers and characters in texts, whose views are ignored and muted because of their sexual orientation, ethnicity, social position, and religious beliefs. In that sense, we need respectful reading.

Within the *fale-'o-kāinga*, we are called to read the Bible wisely (*fakapotopoto*). Very often our academic aggressiveness produces unwanted results, because we make the task of interpretation so complicated and therefore rob the Bible from the hands of those with whom we are trying to read. We need to manage the way we read wisely; we need wise readings.

Within the *fale-'o-kāinga*, we are urged to give other ways and views a chance and allow more participation in the process of interpretation. We are also required to critique any interpretation or any text that claims to have the right answers. We need participatory reading.

Within the *fale-'o-kāinga*, we are encouraged to read responsibly. We have to show through the way we interpret texts that we care; we have to be responsible readers and ask ourselves to what end is our reading and whose interests are we serving. Biblical interpretation needs responsible readings.

Finally, within the *fale-'o-kāinga*, interpretation has to be done with epistemic humility. We need humble readers. That involves readiness to admit one's limitations, to acknowledge the contribution of others, and to be open to criticism and change when necessary.

These guidelines are proposed with hope that they may initiate an epistemic shift from complex and abstract theories that have guided academic research for so long to familiar and concrete insights that are derived from lived experiences of real people in their contexts. The permission for such a shift comes from the understanding that interpretation of any kind is situated within, and shaped by, one's cultures and contexts.

Closing Remarks

We need to rethink (and rewrite) our sense of belonging again and again from our contexts and utilize the resources that we have at our disposal, for if we do not do it ourselves, somebody else is going to do it for us (see Hau'ofa 1993). Let us mind/mine the way we read within the *fale* that we share and acknowledge that we are all part of a *kāinga*, a household, a family.

We participate in biblical interpretation as diverse people with different ideas, cultures, and interests. But let us not forget that we do so as a *kāinga*, who belong to one global *fale* (ecology/economy), and because we can no longer isolate ourselves from each other, we must learn somehow to read the Bible *respectfully, wisely, participatorily, responsibly, sincerely,* and *humbly* together.

Works Cited

Hau'ofa, Epeli. 1993. "Beginnings." Pages 2–16 in *A New Oceania: Rediscovering Our Sea of Islands.* Edited by Eric Waddell, Vijay Naidu, and Epeli Hau'ofa. Suva, Fiji: University of the South Pacific.

West, Cornel. 1993. *Keeping Faith: Philosophy and Race in America.* London: Routledge.

Celebrating Hybridity in Island Bibles: Jesus, the *Tamaalepō* (Child of the Dark) in Mataio 1:18–26

Mosese Ma'ilo

The experimental translation of island Bibles in the nineteenth century was controlled by Western missionaries (esp. London Missionary Society and Wesleyan missionaries).[1] But their superior attitudes towards native tongues were not hegemonic enough to guard what biblical scholars refer to as the "originality" of biblical languages. Hebrew and Greek—with their associated cultural symbols—were not universal enough to remain unaffected when crossing the barriers of language and cultural difference. Missionary translators could not resist the pressure of island (recipient) languages in order to effectively transfer biblical ideas to island readers. Likewise, island languages were fairly limited to fully accommodate biblical ideas. The cultural politics of "difference" is inescapable in any textual translation, and the Bible is of no exception. A number of examples from island Bibles indicate the "ambivalent" nature of the Bible translation process. The poetics of such Bibles is and remains neither the one (Greek/Hebrew) nor the other (Samoan/Fijian/Tongan, etc.). Any claim on either is perhaps untenable.[2]

---

1. Oceania Bibles were produced during the era of British and European imperial expansion in the South Seas and all over the world. Missionary translators were agents of such imperial expansion in terms of exporting Victorian Christianity.

2. The ambivalence of native Bible languages does not mean that indigenous Bibles did not accomplish the desire for Christian religious univocity or the movement towards planting Christianity in Oceania. It was indeed accomplished, but just like the ambivalence of cultural representation, it was an achievement within the ambivalence of mimicry, a type of Christianity that was neither completely Western nor totally indigenous. It was universal Christianity not in terms of sameness or difference, but in

Nevertheless, there is a touch of value to be celebrated with island Bibles. As literary productions of imperialism, they could be treated as postcolonial texts. Their introduction, reception, and translation indicate the colonial significations of cultures and languages, based on the notion of "difference," of otherness. As a result, island Bibles constitute a poetics of imperialism; languages packed with Western-oriented Christian culture. The desire of missionary translators was not entirely to translate God's word. It has to be a word with the power to convert, dominate, and redirect the "savage" islanders' moral, spiritual, and cultural consciousness through the power of language, *island Bible languages*. Consequently, island readers were compelled to abandon and demonize their own native dialects, together with their associated symbols and worldviews, in favor of Western and biblical values. This is articulated in the following sections by analyzing the Samoan translation of Matt 1:18–26, thereby calling attention to the politics of translation and the gifts of native languages.

1. Island Bibles: Neither the One nor the Other

How can we emancipate island readers from the grip of Western ideas, embedded within the conceptual framework of island Bibles? Certainly, retranslation or revision is reasonable. However, postcolonial theory suggests a rather challenging way that both critiques and respects the colonial past. Island Bibles exemplify hybridity—neither the one nor the other. Hence, they become authentic resources for doing island biblical interpretation. Let us turn to some of the postcolonial suggestions.

Samia Mehrez, a postcolonial literary critic refers to these types of texts (after translation) as hybrids, since they are culturolinguistic layered.

> These postcolonial texts frequently referred to as "hybrid" or "métissés" because of the culturo-linguistic layering which exists within them, have succeeded in forging a new language that defies the very notion of a "foreign" text that can be readily translatable into another language. (Mehrez 1992, 121)

The language of a text after translation is "new" in the sense that it has its own distinctive linguistic and cultural character. While translation has, in

---

terms of hybridity, an offshoot of both Western Christianity and Oceania indigenous religious and cultural experiences.

a sense, forged a new language, island Bible languages are multilinguistic and multicultural enough to resist the idea of a dominant cultural perception, monopolized by the values of a single culture. Mehrez's observations of the form and literary nature of these texts is obliging to place some value on island Bible translations, which are always referred to as merely copies, whose true meanings are controlled by Greek or Hebrew original texts.

In referring to Bible translation, Homi Bhabha (a leader in postcolonial thinking) points out the ambivalent nature of the translating process. In the mission context, the act of translation relocated the Bible from being an insignia of colonial authority to becoming a tool for resistance; from being a fetish to becoming a hybrid. Translation, then, allowed the written authority of the Bible to be confronted and defied in the postcolonial context.

> The process of translation is the opening up of another contentious political and cultural site at the heart of colonial representation. Here the word of divine authority is deeply flawed by the assertion of the indigenous sign, and in the very practice of domination the language of the master becomes hybrid—neither the one thing nor the other.... The written authority of the Bible was challenged and together with it a postenlightenment notion of the "evidence of Christianity" and its historical priority, which was central to evangelical colonialism. (Bhabha 1994, 33–34)

Bhabha's use of "hybrid" to designate the master's language after translation points out the possibility of new ways to enliven the dialogue on methods and resources for biblical interpretation. As hybrids, island Bible languages become an unexpected sign of resistance in reading and interpretation. Bhabha argues that if translating the Bible into the *other's* language has opened up a combative site for colonial representation, it is upon that same political site (for instance, the Tongan, Tahitian, or Samoan Bible) where authoritative readings could be challenged. If translation (because of the assertion of the indigenous sign) robs the word of divine authority, then the reassertion of the indigenous sign—the island Bible—argues against the colonial authority of biblical languages. Island Bibles represent a language *in-between* that occupies what Bhabha terms the "Third Space" of negotiation: a space where foreign and local symbols are brought into the harmonious creation of new concepts and meanings (36–37).

2. Island Bibles: Fresh Lifeblood of the Christian Manual

From the point of view of cultural translation studies, the language of a translated text is the continued life of a dead text in another life context. Susan Bassnett argues that a translation is

> the continued life of a text at another moment in time.... translation therefore becomes the act that ensures the life of the text and guarantees its survival.... [A translation] injects new lifeblood into a text by bringing it to the attention of a new world of readers in a different language. (Bassnett 1996, 22)

To reread Bassnett's view for our purpose, island Bibles are not merely translations for the sake of translation. They are the continued life of the Greek and Hebrew texts at another moment in time. They establish the ongoing life of the Bible, the manual of Christianity, in island languages and cultures. Island Bibles inject new lifeblood, island lifeblood, into dead Hebrew and Greek languages. Island Bibles are texts where the foreign God speaks directly and originally to the real world of island readers, no longer as a foreign God, but as a universal God that speaks in a tongue that is understood by island readers, while remaining part of a universal Christian language. Our failure to treat island Bibles as equal authority to the so-called original texts indicate neocolonial attitudes, where interpreters of original texts or the more sophisticated Western Bibles think, "Thanks to translation, we become aware that our neighbours do not speak and think as we do" (Paz 1992, 154).

The hybrid identities of island Bibles do not signify cultural authority or any claim to language superiority. Their languages "remain perpetually in motion, pursuing errant and unpredictable routes, open to change and reinscription" (McLeod 2000, 219). Biblical studies in our postmodern and postcolonial island contexts need this perpetual movement allowing diverse cultural experiences to pursue unpredictable readings and interpretations. The experiences of island biblical scholars offer alternatives. Taking island Bibles seriously as God's word on the same *kerygmatic* caliber (as with the original intention of the Hebrew and Greek texts) may offer a challenge and injects new lifeblood into the future of biblical studies in the islands.

3. Mataio 1:18–25: Jesus as *Tamaalepō* in Samoan Bible

Through Bible translation, we have in our possession the biblical resources for island biblical interpretation. Below is an application of such theory to a rereading of Mary's pregnancy narrative in Mataio 1:18–25 (Samoan Bible).

Samoan Translation (1887)
[18] Sa faapea ona fanau mai o Iesu Keriso. Sa faufautane lona tina o Maria ia Iosefa, ua iloa ua *to* o ia i le Agaga Paia, a o lei faatasi i laua. [19] O Iosefa foi lana tane o le tagata amiotonu ia, e lei loto foi o ia ina faamasiasi ia te ia i luma o tagata, ua ia manatu e faatea lemu ia te ia. [20] A o manatunatu o ia i na mea, faauta, ua faaalia mai ia te ia agelu a le Alii i le miti, ua faapea mai, Iosefa e, le atalii o Tavita, aua e te fefe ina aumai ia te oe o Maria lau ava; aua o lana *to* mai le Agaga Paia lea. [21] E fanau mai e ia le tama tane, e te faaigoa foi ia te ia o Iesu; aua e faaola e ia lona nuu ai a latou agasala. [22] Ua oo nei mea uma ina ia taunuu ai afioga a le Alii I le perofeta, ua faapea mai, [23] Faauta, e *to* le taupou, ma fanau mai le tama tane, latou te faaigoa foi ia te ia o Emanuelu; o lona uiga pe a fa'amatalaina, Ua ia te I tatou le Atua. [24] Ua ala Iosefa, sa moe, ona faia lea e ia pei ona fai mai ai ia te ia o le agelu a le Alii; ua na aumai lana ava. [25] Ae la te lei feiloa'i ua oo ina fanau mai e ia o lana tama tane ulumatua; ua ia faaigoa foi ia te ia o Iesu.

NRSV
[18] Now the birth of Jesus the Messiah took place in this way. When his mother Mary had been engaged to Joseph, but before they lived together, she was found to be *with child* from the Holy Spirit. [19] Her husband Joseph, being a righteous man and unwilling to expose her to public disgrace, planned to dismiss her quietly. [20] But just when he had resolved to do this, an angel of the Lord appeared to him in a dream and said, "Joseph, son of David, do not be afraid to take Mary as your wife, for the child *conceived in her* is from the Holy Spirit. [21] She will bear a son, and you are to name him Jesus, for he will save his people from their sins." [22] All this took place to fulfill what had been spoken by the Lord through the prophet: [23] "Look, the virgin shall *conceive* and bear a son, and they shall name him Emmanuel," which means, "God is with us." [24] When Joseph awoke from sleep, he did as the angel of the Lord commanded him; he took her as his wife, [25] but had no marital relations with her until she had borne a son; and he named him Jesus.

3.1. Translation as Representation

From a translator's point of view, the rendering of Mary's "pregnancy" (Matt 1:18, 20, and 23) in the Samoan Bible reflects one of the pathetic strategies employed in missionary Bible translating in the islands. In most cases, like this one, they opt for simplicity rather than the most appropriate terms available in island tongues. With reference to missionary representations of islanders, the option for simplicity signified how islanders were characterized in missionary writings as simple and stupid: simple people, simple minds, simple language.

3.2. Situation of Maria's *To* (Pregnancy)

Matthew does not directly address the situation of Mary as "pregnant." The direct term (for pregnancy) only appears in Luke 1:24, with reference to Elizabeth's condition. Instead, Matthew uses an expression that literally means "with child." The weight is on the state of the mother's union with her child, although the "how" did she become pregnant is always present without proper explanation.

Translators of the Samoan Bible render "with child" into *to*,[3] which is the simplest and greenest term for pregnancy. They ignored other appropriate and more respectful terms in the Samoan idiom such as *tau'ave le tama*, meaning "to bear about" (Pratt 1911, 297) or "to carry," which best serves Matthew's "with child." The term *ma'itaga* (192: "a confinement") is another respectful term for pregnancy. Nevertheless, our study takes and respects the translator's decision as a hybrid, neither the one nor the other.

Maria (Mary) was pregnant before she and Iosefa (Joseph) consummated their relationship (Mataio 1:18). The most appropriate (Samoan) term for such premarital pregnancy is *toifale* or *tofale* (Milner 1993, 268), which literally refers to "pregnancy outside of marriage." Such is a "disgrace" and humiliation that a girl brings to her family and thus the child is pronounced illegitimate. Translators avoided *tofale*, perhaps because Maria's situation is exceptional (divine plan?). But the avoidance of *tofale*

---

3. George Pratt (1911) defines *to* as "with child" in the Samoan-English part of his Dictionary. In the English-Samoan part of the same Dictionary, "pregnant" is defined as *to*, with *alo* as the respective term for a chief's wife. Pratt is actually playing with words here, to comply with their uses in the Bible. G. B. Milner (1993) defines *to* as "be pregnant, pregnancy."

confuses the poetics of the passage. The content and context of the passage says it is *tofale*, but the term *to* allows otherwise. Interestingly, *to* silently leaves the situation opened for Samoan readers to be branded as *tofale* or otherwise. But since the passage has the authority to overrule the meaning of terms, Maria's pregnancy in the Samoan Bible is unfortunately, *tofale*.

3.3. Maria the *Taupou* (Virgin) Mother?

In Mataio 1:23, the rendering of virgin into *taupou*[4] to designate Maria as sexually pure raises a concern. *Taupou* is a village title, an honorary designation given to a special lady not so much because of virginity, but of status and responsibility. *Taupou* is always the daughter of a village high chief. She is responsible for the maidens of the village and women's organization, the *aualuma*. She can or cannot be a virgin, as her status as *taupou* is not to be determined by her virginity, but by her ancestral line.

The use of *taupou* really brings the text to the Samoan sociopolitical context. It portrays Maria as a responsible and respected woman leader in the Samoan society, rather than a Jewish virgin in the sense of sexual purity. Based on such reference, Maria in the Samoan Bible language can and cannot be a virgin mother. She is a *taupou*, a respected and responsible woman leader. Her title is not based on her sexual purity, but on status and responsibility in a community. This also indicates the richness of hybridity in translation.

3.3.1. The Identity of Maria's Child in Mataio 1:18–25

Maria's premarital pregnancy or *tofale* in Mataio 1:18–25 means that the child she gives birth to is "fatherless." Andries van Aarde (2002, 71) argues for the same identity of Jesus from a psychohistorical portrayal of Joseph. Van Aarde contends that Jesus is fatherless, because the Joseph of the gospel writers is an ideal type, and that he never had sex with Mary. By reconstructing the life of Jesus within first century Herodian Palestine, van Aarde's ideal typology of Jesus's life indicates that he lived a life of defending fatherless children, patriarchless women, and other social mis-

---

4. George Pratt (1911, 303) translates this "title" distastefully as "a virgin." But Milner (1993, 255) defines *taupou* appropriately as "title of village maiden (a position held according to Samoan custom by a virgin singled out for her charm, looks, and manners. Among her duties is the preparation of kava" (see also Kramer 1994, 34–37).

fits—like Jesus himself. His relationship to his family and the leaders of the society within the Gospels indicates that Jesus was treated as a fatherless child until God took him as his child.

Mark, the first gospel to be written, makes no reference that Jesus has a father. Jesus is referred to as the carpenter, the son of Mary (Mark 6: 3). In the Jewish tradition, to call a man the "son of a woman" means that his paternity is questionable. John Spong (1996, 205) rightly argues on this matter with reference to the early church traditions: "At some point after Christians fought off charges based on the scandal of the cross, they clearly had to fight off charges based on the scandal of the birth."

Samoan Bible readers and interpreters know this scandal by reading the Bible in their own tongue. A fatherless child, like Maria's son Jesus, is thus identified as the *tamaalepō*, a "child of the dark" (Milner 1993, 185). The child's father is the night, the darkness, which is actually a discreditable social identification. *Tamaalepō* is an illegitimate child because of the mother's "having sex in the dark," a symbolic expression of pregnancy as a result of sex outside of marriage. Not only is the child fatherless, the child is also a *tamaalepō*, based on the way he or she was born into the family. Following van Aarde's archetypal reconstruction of his fatherless life, Jesus did not enjoy the privileges of a legitimate child with a father at his side. This is in line with a *tamaalepō* in the Samoan society, who is expected to endure the humiliation and disgrace in family and society.

3.3.2. From *Tamaalepō* into *Tamaaāiga*: Emancipative Interpretation

Tamaaāiga is a person of large family connections (Pratt 1911, 318).[5] From a Samoan perspective, Jesus's genealogy in Mataio 1:1–14, narrated before the virgin birth, states that he is a person of large family connections. Jesus is related to the nobilities and fathers of the nation, so he is also a *tamaaāiga*. Jesus was perceived as *tamaalepō* by the members of his society, which means "being barred from status as child of Abraham, that is, a child of God" (van Aarde 2002, 81). He was treated as such in the Gospels. Even his own family were standing far from him (Mataio 12:36–50). He

---

5. *Tamaaāiga* refers to the paramount chiefs of the Samoan society. There is tradition that most *tamaaāiga* were initially *tamaalepō* or children of the dark.

faced Roman persecution, because of his beliefs in how a fatherless child becomes a child of Abraham.

Nevertheless, it is the memory of the faith community, after the resurrection, which transformed the identity of Jesus from *tamaalepō* into *tamaaāiga* based on his life, his teachings, and how he ended his life as a true child of God. The changing identity of Jesus is a result of his victorious death. He died a *tamaalepō* but was raised to become a *tamaaāiga*. When Matthew starts his story of Jesus with a genealogy, it is obviously with the purpose of confirming Jesus's identity as *tamaaāiga*, a person with large family connections, beginning with Abraham, David, and the nobility of the chosen nation.

This reading serves the island community better, by reading from their own Bible. It is an emancipative reading of Mataio 1:18–25, which hopes to transform island cultural perceptions of *tofale* women like Maria and the *tamaalepō* (or children of the dark) like Jesus. Today, although islands are Christianized, these social humiliations still persist and result in either young people leaving the family for good or even suicide.

We islanders need to come up with sound biblical interpretations that are not only meaningful, but have the power to transform island perceptions of life in the modern world. The way forward is not by absolutizing our island cultural values in biblical interpretations, but by mobilizing their transformative aspects with the relevant biblical texts for emancipative reading. That is, reading that helps transform our own oppressive, despotic and patronizing perceptions.

4. Concerns and the Way Forward

The hope of this type of reading is clear: to resignify island Bibles not only as texts for mission and conversion (as they were used in colonial evangelical mission), but as textual resources for sound and emancipative biblical studies. In every Oceania theological institution, island Bibles are treated as translated copies of the Hebrew or Greek originals. In terms of proper biblical studies, they are never equal to the authority of original texts (and sometimes the English versions) simply because they are "translations." This attitude shows the ranking of Bibles, which raises a grave concern with regard to categorizing the inspired word of God according to cultural/linguistic difference. Island interpreters (in academia) are forced to depend on Western commentaries, consensus of (male/white/European) biblical scholars, and European guilds of biblical scholarship for standardized biblical interpretation and methodology.

5. Island Bibles and Vernacular Biblical Hermeneutics

In *Vernacular Hermeneutics*, R. S. Sugirtharajah (1999, 98–105) demonstrates three modes of vernacular reading: the conceptual correspondences, the narrative enrichments, and the performative parallels. The first mode "seeks textual or conceptual parallels between Biblical texts and the textual or conceptual traditions of one's own culture," looking "beyond the Judaic or Greco-Roman contexts of the biblical narratives, and seeks corresponding conceptual analogies in the reader's own textual traditions." The second mode looks at reemploying "popular folk tales, legends, riddles, plays, proverbs, and poems that are part of the common heritage of the people, and place them vividly alongside biblical materials, in order to bring out their hermeneutical implications." The third mode has to do with the utilization of "ritual and behavioural practices that are commonly available in a culture." All three modes are about the revitalization of indigenous texts, cultural, popular, and ritual values. As Sugirtharajah himself acknowledges at the end, there are weaknesses as well as positive implications of formulating vernacular biblical hermeneutics this way.

In my view, most of the problems in island biblical hermeneutics so far are related to the question, "Where do we locate culture or native traditions in the hermeneutical activity?" In building on both the negative and positive sides, I suggest that the limitation we encounter is based on the fact that we "locate" culture as the source of the interpretative conflict rather than as the effect of discriminatory practices. Cultural difference is not the problem, because it was a strategy of inscribing power, control, and authority in the colonial mission context (Bhabha 1994, 114). It means biblical hermeneutics should not be an exercise of recovering island cultural identity. Cultural difference was and is still not the problem. The problem is when other interpretations, worldviews, and cultural perceptions of reality are absolutized by devaluing other people's readings and experiences. Relocating culture as the source of conflict in biblical interpretation implies that we, island readers, are reinscribing discriminating practices that we seriously tend to decry.

Island biblical hermeneutics has to kick start by acknowledging which Bible "telling" or Bible language we are reading from. It is not about the absolutization of our island Bibles as a mark of identity, but of acknowledging and celebrating their hybridity, neither the one nor the other. The idea is to take the hybridity of island Bible languages as a site of resistant (to the dominant readings) biblical interpretation. It is "biblical" in terms of respecting the poetics of island Bibles as Bibles, where cultural variables are to be resignified as "referents" whether they are conceptual correspondences, narrative enrichments, or performative parallels.

While cultural or native elements are treated as referents, island biblical hermeneutics is in a position to avoid the absolutization of island cultures. Sugirtharajah cautions,

> At a time when vernacular cultures and languages are intermingled with those of the metropolis, it is not always feasible to use dialect as a test of identity. In our enthusiasm to recover the native, we may put ourselves in the double predicament of finding redeeming values both in the indigene and in the text.... By eulogizing the ascendancy of the native and revalorising the text, we may end up by fixing, absolutizing and immobilizing both. (Sugirtharajah 1999, 15)

It would be naïve if the purpose of island biblical hermeneutics is to eulogize the native and his or her cultural values as absolute. More confidently, I assert that the purpose of island hermeneutics is to "uncover" island presence in a postmodern and postcolonial world. In that sense, island biblical hermeneutics indirectly presents the native's presence as dynamic in biblical interpretation. Island interpretative capabilities have long been overshadowed and marginalized by dominant Western, Latin American, African, and Asian worldviews. Island hermeneutics resignifies island readers' presence, voices, and memories in biblical interpretation.

The point is that island biblical hermeneutics must focus on the present reality of their own Bible's linguistic and cultural hybridity. There is a difference between recovering the islander's culture and recovering the islander as a hybrid cultural being. The former is perhaps plausible if any pure indigenous culture still exists. The latter resignifies our experiences, memory, and presence as not culturally exclusive. Therefore, island biblical hermeneutics must take the native Bible seriously. This should *not* be based on island Bible languages as mark of indigenous identity, but of their hybridity. Starting with island Bible languages also builds the confidence of island readers to express their biblical expositions based on how they read the Bible in a familiar tongue. At the same time, we are also aware that such interpretation is part of a global biblical dialogue, where nothing is lost, but gained in translation.

Works Cited

Bassnett, Susan. 1996. "The Meek or the Mighty: Reappraising the Role of the Translator." Pages 10–24 in *Translation, Power, Subversion*. Edited by Román Álvarez and M. Carmen-África Vidal. Topics in Translation 8. Clevedon: Multilingual Matters.
Bhabha, Homi K. 1994. *The Location of Culture*. London: Routledge.
Kramer, Augustin. 1994. *The Samoa Islands: An Outline of a Monograph with Particular Consideration of German Samoa*. Translated by Theodore Verhaaren. Vol. 1. Honolulu: University of Hawaii Press.
McLeod, John. 2000. *Beginning Postcolonialism*. Manchester: Manchester University Press.
Mehrez, Samia. 1992. "Translation and the Postcolonial Experience: The Francophone North African Text." Pages 120–38 in *Rethinking Translation: Discourse, Subjectivity, Ideology*. London: Routledge.
Milner, G.B. 1993. *Samoan Dictionary*. Auckland, NZ: Polynesian Press.
Paz, Octavio. 1992. "Translation: Literature and Letters." Translated by Irene del Corral. Pages 152–62 in *Theories of Translation: An Anthology of Essays from Dryden to Derrida*. Edited by Rainer Schulte and John Biguenet. Chicago: University of Chicago Press.
Pratt, George. 1911. *Pratt's Grammar and Dictionary of the Samoan Language*. 4th ed. Malua, Samoa: Malua Printing Press.
Spong, John Shelby. 1996. *Liberating the Gospels: Reading the Bible with Jewish Eyes*. San Francisco: HarperCollins.
Sugirtharajah, R. S. 1999. "Thinking about Vernacular Hermeneutics Sitting in a Metropolitan Study." Pages 98–105 in *Vernacular Hermeneutics*. Edited by R. S. Sugirtharajah. The Bible and Postcolonialism 2. Sheffield: Sheffield Academic Press.
Van Aarde, Andries. 2002. "Jesus as Fatherless Child." Pages 65–84 in *The Social Setting of Jesus and the Gospels*. Edited by Wolfgang Stegeman, Bruce J. Malina, and Gerd Theissen. Minneapolis: Fortress.

Creolizing Hermeneutics:
A Caribbean Invitation

Althea Spencer Miller

At the 2012 Society of Biblical Literature Annual Meeting, Fernando Segovia,[1] in a panel presentation for theorizing islandedness, suggested availing the intellectual history of islands for the development of island hermeneutics. The Négritude Movement as a partial Caribbean intellectual movement appeared an obvious recourse.[2] Two of its more famed luminaries were Frantz Fanon and his teacher Aimé Césaire, both from Martinique. A lesser-known but inspiring later luminary was Édouard Glissant (Martiniqian novelist, poet, playwright, cultural, and narrative critic). This chapter explores aspects of Glissant's thought in conjunction with a sense of Caribbean islandedness. It presents islandedness as evidentiary and constructive for a Caribbean biblical hermeneutic. It does so by interweaving anecdotes with reflections on physical islands as organic metaphoric realities and analyzes Glissant's thoughts on creolization, history, and language/orality as a sample of the Caribbean region's intellectual history.

---

1. Fernando Segovia, well known for his contributions to New Testament studies as a postcolonial critic spent much of his early life in Cuba and is referenced here as a Caribbean islander.

2. Négritude is an aesthetic and ideological movement that emerged among francophone black intellectuals in the 1920s and 1930s, including Aimé Césaire (Martinique), Léon-Gontran Damas (French Guiana/Martinique), and Léopold Sédar Senghor (Senegal), all of whom met in Paris as students. The three, whom Frantz Fanon later joined, were concerned with the establishment of "blackness" as a way of resisting French colonialism and racism. Little known is the role of sisters Paulette and Jane Nardal who were instrumental in providing a meeting place for the students and connecting them to the Harlem Renaissance. These were the proponents of the movement. Dissenters include Wole Soyinka, who thought the movement was misguided in its assertions of black pride and had some accomodationist tendencies in their race ideology.

The method focalizes Caribbean lived experience as an acceptable origination for contributions to a transnational conversation.[3] Community life, ancestral memories, and colonial history ground the anecdotes as cultural roots for this particular hermeneutic. The anecdotes include an induction service held in a Methodist church in Saint Thomas, Jamaica, a speech by Dr. M. Douglas Meeks at the Oxford Institute of Methodist Theological Studies, and a conversation between a congregant and pastor in Kingston, Jamaica. Altogether, the anecdotes raise issues of creolization, history, and language, each of which signals the mix of carnage and opportunity that is the legacy of colonialism and its reinforcement by a superimposing Eurocentric propagation and practice of theology. The anecdotal beginning is, implicitly, a rebuttal of Caribbean theological and hermeneutical Eurocentrism and a reach for the eclectic authenticity of the Caribbean cultural experience.

The islander approach has decolonizing intent to unfetter hidebound commitments to externally-oriented and heterogeneous theological frameworks. I adopt a liberatory approach that resists termini and definitive frames. Rather, I explore and seek a hermeneutical direction. Édouard Glissant's *Caribbean Discourse: Selected Essays* (1996) is the primary interlocutor toward that end. With islander openness to oceanic expanses and invitations, this paper's conclusion will be open also. Additionally, nonconclusion is a methodological nod to an anticipated conversation. That is, it expresses a commitment to a communal rather than an idiosyncratic and individualistic claim to sagacity, regional or otherwise. This contribution to Caribbean biblical hermeneutics opens to a transnational biblical conversation.

African Insinuations, Christian Inarticulations, and Creolization

The induction service of a probation minister into a Methodist circuit in Jamaica had been observed in proper English liturgical fashion. Afterward, clergy and laity headed toward a small church in the hills for a communal repast. The setting was memorable, exciting, and delightful. Tables were set with banana leaves. Crockery included calabash gourds for drinking

---

3. Édouard Glissant often used the term global consciousness when criticizing Eurocentric historicism but also when forespeaking the transformative potential of creolization. At times, I use the decentering and decolonizing term "transnational" instead.

hot chocolate made from unprocessed cocoa sticks. Bottle lamps[4] cast a flickering glow over the proceedings as the darkness thickened. The feast included run down,[5] green bananas, and tightly rolled cornmeal cartwheel dumplings. I relished this almost pristine presentation of our Jamaican culinary and agricultural heritage with the accentuating natural table service. Here was an elevation of the food of the poor from the ground and a recognition of the celebratory function of Jamaican products. The entertainment was yet to come. When it did, I experienced a moment of unforgettable self-illumination.

The drummers' performance provided the energy for the entertainment. Drummers played conga drums using a complex interplay of rhythms that differed from both Jamaica's popular music forms and even more so from the liturgical music of the church service. These unfamiliar rhythms evoked unfamiliar stirrings within me, with primeval and ancestral rawness. Deep recesses of my being quivered with rising echoes of an African past. The forsaken psychical memories of my African foreparents roused themselves as evocations of a people I had never known and a land I had never seen. I was awash with a strengthening urge to dance when harsh reality confronted these emerging instincts. I did not know how to dance to the drums. There were no accompanying muscular twitches. An unbridgeable chasm and a deeply ingrained resistance to that memory separated my body and psyche. My clearest and most lingering insight was that European enculturation through education had provided supplanting rhythms. For a moment, the two poles of my interiority stared aghast at each other before, inexorably, the chasm closed.

The church service and the communal repast framed that experience and clarified the culturally polarized textures of the event. The former had the cast of approbation, accustomedness, and oversight. The latter emerged from a section of Jamaican culture that contains and expresses the memories that will not die. The liturgical exercise was standard traditional British Methodism. The communal repast evoked memories that

---

4. Bottle lamps were usually made from tall, slim soda bottles containing kerosene oil, a flammable liquid. A bottle length paper wick steeped in kerosene oil stuffed the bottle's mouth. Thus, the light would burn continuously.

5. Run Down (a.k.a. Dip and Fall Back) is an old Jamaican recipe made with mackerel, seasonings, and coconut milk, which forms a thick, almost custard-like sauce. It is served with boiled green bananas and dumplings. It might also be served with breadfruit or Jamaican yam.

had no place in the liturgy. Those vestiges, by struggling to be more than reminiscences, indicated their persistence. I was an anglicized being with vestiges of another cultural reality that I would do well to know: a creolized[6] person.

Creolization in the Caribbean is a product of colonization. It is akin to Dubois's (1903, 1–7) "double consciousness," with a less obvious veil. The Duboisian veil has an external reality that produces and nourishes an interiorized sense of separation from the dominant society. That veil has institutional, societal, cultural, and relational mores and forms that define and demarcate the contours of the veil. In the postcolonial/neocolonial world of 1980s Jamaica, there was no such external veil for a middle class child with a middle class future (see Walcott 2005, 257–64).[7] A hierarchy of cultures that replicated the hierarchies of Jamaican society with its prioritization of Eurocentrism constituted the Jamaican creolized interiority. The veil was invisible and internalized. Glissant detects a similar invisibility in creolization. Paradoxically, he characterizes creolization as "the unknown awareness of the creolized" (Glissant 1996, 2). Insofar as there is a veil in Jamaican society, the intersections of race, acculturation, and class fuzz the veil. Further complications are evident through the alignment of cultivated Eurocentrism with elitism and the corollary alignment of African retentions with subaltern classes. This might be an even more deeply damaging and destructive veiling. But there is something more. Glissant envisages creolization disrupting History. History (with an uppercase H) is the invention and servant of developing European nationhood and colonial self-medication. It is a

---

6. Édouard Glissant used the term "Creole" in his discussions of the interiority and language of Martiniqans. He applied it with different valences. He saw Creole as a language to be as deformed as French was for Martiniqans. It was therefore useless for Martiniqan development. With fervent pessimism, he deemed it a language doomed for extinction. Creolization of the psyche, in his thinking, boded much more promise for the creativity and generativity of Martiniqans. I am using "creolization" as applied more broadly to the culture and psyche rather than to Creole as a hybrid of French. See Michael Dash's (1996, xxii–xxiii) insightful discussion of Glissant.

7. In a response to what he deems as V. S. Naipaul's insolent appraisal of Caribbean life as mimicry, Walcott reabsorbs mimicry into human and historical genetics. Walcott's essay is a strident critique of and response to V. S. Naipaul's *The Mimic Men: A Novel* and locates mimicry as inevitable within the geographical proximities of the Americas. In doing so, he approximates the Duboisian notion of the veil in terms of regional relationalities rather than in terms of intranational relationships.

totalizing and homogenizing force. Creolization diffuses, defuses, and (re)fuses the "'totalizing' pretensions of History" (Dash 1996, xxviii). Glissant is blistering in his criticism of History.[8] Dash summarizes Glissant's critique as follows:

> Because no truly total history (in all its diversity) is possible, what History attempts to do is to fix reality in terms of a rigid, hierarchical discourse. In order to keep the unintelligible real of historical diversity at bay, History as system attempts to systematize the world through ethnocultural hierarchy and chronological progression. Consequently, a predictable narrative is established, with a beginning, middle, and end. History then becomes, because of this almost theological Trinitarian structure, providential fable or salvational myth.... History ultimately emerges as a fantasy peculiar to the Western imagination in its pursuit of a discourse that legitimizes its power and condemns other cultures to the periphery. (Dash 1996, xxxix)

Creolization declares tidy History to be a falsification and a disfiguration of the actuality of historical existence among peoples relegated to ahistory and prehistory. Creolized humanity collectively indexes the fissures in History's corralled march for European ethnocentric dominance. The drummer, the drums, the dance, the conflict between European cultural impositions and repressed African memory, the impulse, the twitch, all declare, "We remember and we are here." In tribute to Glissant's vision, I affirm and aver a ceaseless creolization, "the synchronic relations within and across cultures that matter more than the rigid diachrony of orthodox historicism ... replac[ing] the falsifying symmetry of history as linear

---

8. Glissant (2001, first written in French) explores the contours, meanderings, and alternate resourcing of narrating history. This brilliant exploration unfolds in an encounter between an old trained archivist and a young *quimboiseur* (healer). Their confluences articulate in novelistic form, Glissant's vision of ceaseless creolization as the disruption of History. Even so, often omitted from Négritude narratives are Martiniqan women's efforts in the 1940s. The Nardal sisters (Paulette, Jane, and Andrée) were leading female thinkers associated with black diasporic journals such as *La Dépêche Africaine* and *La Revue du monde noir*. Paulette Nardal, in particular, was renowned for the journal *Woman in the City*. Jane Nardal "described [Jane's] coming into race consciousness to Alain Locke as 'An autobiography of a Re-Colored Woman'" (Nardal 2009, 2). *Beyond Negritude* (Nardal 2009), a monograph on women's contribution to the Négritude movement, contains a brief introduction to Paulette Nardal and includes notes on other Martiniqan women involved in critical race analysis.

progression" (Dash 1996, xxix). Ceaseless creolization ensures the failure of totalizing, systematized, linear progressive, Eurocentric History. Creolization is messy and filled with ruptures, options, alternatives, multiplicities, contradictions, inventions, and creativity.

Because of an ironic colonial machination, of persistent political, economic, and cultural colonization, Martiniqans are a study in ceaseless creolization. However, Glissant, in *Caribbean Discourse* seems pessimistic about the fate of Martiniqans and with good reason. In 2015, Martinique, having produced Aimé Césaire, Franz Fanon, and Édouard Glissant, occupies the status of a department in France's government.[9] As Dash 1996, xli) observes, Glissant presents Martinique "as a solitary and absurd denial of the cross-cultural imagination by its desperate attachment to metropolitan France. Assimilation has meant for overseas departments, like Martinique, a denial of collective memory, of regional identity." Apparently, there are limits to the virtuous impact of ceaseless creolization: fixity cannot characterize creolization. It is a sign in perpetual motion, a study in contrasts and permutations that cannot be static. Creolization produces and is produced by the undulations of unfettered history, real history. Martinique, arguably, is like the Jungian shadow in ceaseless creolization, not oppositional, but an indication and index of an existence that is real and illusionary, a study in hetero/homogeneous, mimicking/inventive paradoxicality.

In the terms of Glissant and other island thinkers, the land is integral to the shaping of consciousness. It is both symbol and reality, both contradiction and consistency, both referent and signifier and all of these simultaneously. On the Caribbean's small islands, there is an in-between ontology that size and location articulate. There is the view of its mountaintops and the loss of its depths in a lapping sea. At each point on its contours,

---

9. By a vote of the French National Assembly in 1946, Martinique ceased to exist as a colony of France. Its new status was the equivalent of governmental departments in Paris. The French colonies of Guadeloupe, Martinique, Guyane, and Réunion also became French departments. Local councils in Guadeloupe and Martinique actively sought departmentalization. Departmentalization reduced the possibilities of future nationhood and non-French cultural development in Martinique. It secured financial dependence on France, the perpetuation of cultural mimesis, and transformed Martinique into a remote governmental outpost. A significant criticism of Aimé Césaire is that he drafted the law that legislated departmentalization. Nick Nesbitt describes departmentalization as bringing "to fruition the ideology of centralization and assimilation" and as creating "an unparalleled degree of dependency upon the metropole." For a fuller presentation of this, see Nesbitt 2003, 7–8.

there is an end that is also a beginning. The road to there ends where the road from there begins. The shoreline that defines land's limits rebuff's ocean's encroachments. This fixed mass in the middle of ocean tides and currents redirects those tides and currents. If the island is small enough, the receding tides accommodate the incoming tides which accommodate the receding tides, all tides swirling to meet new tides at island's tips. The island perpetually disrupts the sea. It weathers storms as the waves wash over, and hurricanes encompass. Then with the calm, the island returns to disrupting the sea, like ceaseless creolization. How else can one read, exegete, interpret, comprehend, apprehend but by demarcation, remarcation, reconfiguration, and then to begin all over again? Glissant again:

> We cannot deny the reality: cultures derived from plantations; insular civilization (where the Caribbean Sea disperses, whereas, for instance, one reckons that an equally civilizing sea, the Mediterranean, had primarily the potential for attraction and concentration); social pyramids with an African or East Indian base and a European peak; languages of compromise; general cultural phenomenon of creolization; pattern of encounter and synthesis; persistence of the African presence; cultivation of sugarcane, corn, and pepper; site where rhythms are combined. (Glissant 1996, xliii–xliv)[10]

This dialectal relationship resists closure and fixity. The veil could be both friend and enemy. Ceaseless creolization is a perpetual journey of undeterminable becoming and an inchoate infinitude. Like the island, it is always beginning but never altogether new.

Relocation in Space and History: Clues from a Historical Recitation

Islands, small islands, tourist Meccas, places of escape, invisible except for illusions of respite and escape, non sequiturs and inconsequential in History's narrative, yet subtle in disruptive irruption. The year was 1983. At the opening plenary of Oxford Institute of Methodist Theological Studies at Oxford University in England, Dr. M. Douglas Meeks, in delivering the keynote address, provided a brief periodic history. The Enlightenment, the

---

10. The ideas of ambiguity, ambivalence, and, here, infinitude were pervasive within the Negritude movement. See Senghor 2005, 184; and Nesbitt 2003, 76–94.

Age of Reason, and Modernity were familiar terms that suddenly seemed strange, alien, forbidding, dangerous, and foreboding. At the end of his speech, Meeks challenged that the practice of Eucharist, instead of being an exclusionary boundary marker, should be a place of radical inclusion.[11] The magnanimity of the gesture was overwhelming, the generosity was capacious, and the intention was genuine. I was captivated and captured. Like a small island in a hurricane suffering the winds' bellow, a force larger than the island, the battering, and the sea defying its boundaries, the whorl of Europe's imprimatur overwhelmed my colonized neurosensors with hope despite its cause: the storm.

Meeks's call to radical inclusion, a quest for prophetic justice, was contingent upon notions of proprietorship, realty rights, and inherent belonging. It was an imperative that reverberates positively among those who receive the challenge as members of the proprietary ranks. Inclusion for the outliers means assimilation. It requires the adoption of someone else's History, the embrace of another's future. It means a loss of language, both *le parole* and *la langue*. This is the process of anticreolization that proposes ethnocide. Radical inclusion implies reenculturation, a radical form of gently coerced historicocultural amnesia, a neoracialization, and baptism that washes away the sin of nonbelonging. The storm roars.

The Enlightenment is a metonym for a period that extends from the end of Medieval Scholasticism to late Modernity. As metonym, it collapses Scholasticism, the Age of Reason, and the development of modern philosophy into one conveyance. It announces the pinnacle of European thought, the rise of rationalism. It pronounces the demise of the mythic, narrative, superstition, and "darkness" of the non-Aryan peoples of the world. Georg Wilhelm Friedrich Hegel, an Enlightenment philosopher, originated the exclusions from history. Aboriginals were prehistory. Africans were ahistorical. Only Europeans were truly historical. Hegel's paradigmatic exclusion of non-European cultures from history reveals the worldview that informed the European encounter with the Tainos and the Caribs of the Caribbean and presages the inhuman(e) treatment of

---

11. At the 2012 Society of Biblical Literature Annual Meeting in Chicago, I had a chance meeting with Dr. Meeks and his wife in an elevator. I shared my memory of this speech with them. His wife affirmed the themes of inclusion as a consistent trait of his thought. I am indebted to that speech. Here, I am more critical than I was then. Without that speech and experience, I would not be positioned now to offer critical thoughts.

African human beings in the slave trade. The Enlightenment is a signifier of the period that Aimé Césaire (2000, 34) denounces as incapable of producing "a single human value" (see also Césaire 1970 and Nesbitt 2003, 118–44). One might say that Césaire Calibanizes the Enlightenment. Inclusion without influence, assimilation with amnesia, neoracialization within ethnocide is the crucible that anticreolization forms. The island that survives the hurricane's passage, if it could speak, might accuse that acquiescence to assimilation is voluntary ethnocide.

Glissant (1996, 5) presages potential Caribbean island theory in his descrier, "From the persistent myth of the paradise islands to the deceptive appearance of overseas departments, it seemed that the French West Indies were destined to be always in an unstable relationship with their own reality … by one of the most pernicious forms of colonization; the one by means of which a community becomes *assimilated*." The mechanism was departmentalization. The effect was racialization. Glissant's declamation that the French Caribbean provided officers and subofficers for the colonization of Africa continues, "They are considered as whites and, alas, behave in that manner." In Africa, the Afro-French Caribbean pseudoelite perform whiteness, and on the island political mimicry institutionalized whiteness in the form of governance structures (8).

However, acquiescence to assimilation is the death knell neither for Martinique nor for the Caribbean. Caribbean islands annually face hurricanes, tropical storms, the rainy season, and droughts. The stable, disruptive stolidity of island as land, as geography, bespeaks resilience, regeneration, and transmutation. To say that the land understands is to assert that the land, the island, answers our questions about history, about aspiration, about dreams, our own, our ancestors, and our children's. The land, the island, the small island informs, shapes, and harbors its peoples' knowing. "So history is spread out beneath this surface, from the mountains to the sea, from north to south, from the forest to the beaches … resistance and denial, entrenchment and endurance, the world beyond and dream. Our landscape is its own monument: its meaning can only be traced on the underside. It is all history" (Glissant 1996, 11). And so Steed Vernyl Davidson (Tobago) turned to the sand, the land, and the sea (in this volume). Leslie James (Grenada) turned to history (James 2012, 11–18).

Ceaseless creolization is a process of negotiating various cultural influences in their relentless presence. Negotiations include adoption, rejection, adaptation, mutation, and transportation with reciprocal impact. In a short essay in the *Africana Bible*, James places colonial experiences of

Caribbean peoples of African descent at the center of history, rethinks, and rearticulates historical periods from that center. He identifies three moments in the brief span of Afro-Caribbean history. They are slavery and the beginning of the African Diaspora, Africana cultural reconfigurations in the "New World," and the third is the Diaspora reconnection with Africa (James 2012, 12–14). Although these are not three discrete periods, James disentangles these three moments in history with anticolonial impact. He utilizes the periodic framing evidenced in Meeks's presentation but imbues it with an alternative historical construction that overlaps and displaces the standard periods of Eurocentric history. James's use of an alternative framing mechanism exposes the pretentiousness of Eurocentric homogenization and systematization. Like the island after the storm, James continues a process of disruption and displacement. One might dare say that James's framing signifies ceaseless creolization. Moreover, his creolization resembles a process that Glissant calls "reversion."

Glissant is uninterested in the distant past as a recoverable memory. He writes, "And there is not one of us who knows what happened in the country over there beyond the ocean, the sea has rolled over us all, even you who see the story and the people in it. There. That is what we call the past. This bottomless sequence of forgetting with, every now and then, some hint flashing into our nothingness" (Glissant 2001, 52). In the narrative situation of a passage in *The Fourth Century: A Novel*, the *quimboiseur* represents the thoughts of a newly arrived slave as assessing the difference between the two coasts. The newly arrived slave understands that "this was no longer the moment to watch for a shoreline, it was time to get ready to live" (49). Living is the goal of reversion. Reversion does not invite a return to an unrecoverable past, nor to origins, nor to an immutable state of being. Reversion invites a return to the points of entanglement. In the James example, the point of entanglement is in the monopoly of history. He catapults from there to theorizing Africana. Reversion is the means by which we avoid death by putting the forces of creolization to work, according to Glissant (1996, 26). The disruptive force of James's historical moments helps to fuel resistance to Eurocentric historical monopoly. It uncovers its pretensions to homogeneity, systematization, and orderliness. It lifts up the historical covers of ahistory and prehistory and says, "Boo!" It is the declarative proof of historical substance in the colonized culture. It peeks into the depths of the seas and sees the ancestral faces of those who died there, enlivens memory, and restores ancestry. It absorbs those unknown, unnamed, and lost and declares them present despite the alien's will to historical obscurantism that haunts History. Glissant's complex relationship to History, James's stalwartness, and the small islands revert, resist, revive, and disrupt. These are heuristic cues. The island paradox announces again that it is precisely within historical incompleteness that the hope of future redemptions and transformations nestle.

THEN THE SUBALTERN SPOKE

Lucy Bailey was an ardent, competent, and exemplary class leader[12] in her congregation in Kingston, Jamaica. In Jamaica's social class structure, her lifetime occupation as a domestic helper marginalized her both socially and religiously. Within her local congregation her character, devotion, constant spoken testimony, and spirituality earned her tremendous respect. On one occasion, her pastor requested that she lead the congregation by offering prayer extemporaneously. She declined. This unexpected, astounding, and perplexing response required an explanation. Her reason was simple. She could not speak well enough. Lucy Bailey was confident beyond any doubt that God heard and answered her private prayers. Why could she not pray in her accustomed speech in a congregation where she was loved and respected? What did it mean to not speak well enough? Did it mean an insufficient command of liturgical English? Did worship as public space delegitimize the language of her holiness? Had Lucy's response betrayed something about her society's marginalization of the *lingua franca* of the Jamaican majority, within our holy two hours?

Patois or Jamaican English[13] is the basic mode of communication for the majority of Jamaicans. Other Caribbean territories have their equivalent. Monolingual use of Jamaican is an indicator of lower social class and educational attainment. There is a similar situation in Martinique. Glissant exclaims over the prejudices and fears of Martiniqan that to speak Creole

---

12. The class system is a way of organizing members in the Methodist church in the Caribbean and the Americas. Members of the congregations are assigned to a class leader who is a trusted layperson. The leader provides a significant measure of pastoral care and oversight to their assignees. Lucy Bailey was a leader at Ebenezer Methodist Church, in Tivoli Gardens, Kingston, Jamaica.

13. Jamaican English is now called Jamaican in recognition of its use as the *lingua franca* of Jamaica. See McFarlane's (1998, 107–21) exploration of Rastafarian pronouns offers insights into the purposeful creolization of language as an alternative epistemology and a response to colonial brainwashing.

alone would be a disadvantage. "It seems that the forces of deculturation no longer need to incite these prejudices. We have all taken over this responsibility" (Glissant 1996, 182). How egregious that the stigma finds approbation in liturgical language. The Lucy Bailey anecdote indicates the need to analyze the relationship between secular speak and formal holy speak. The sociocultural impact of liturgical linguistics aligns it with intracultural and personal dispossessions. The linguistic taboo is an indication of the diminution of the creolizing capacities of bilingual[14] Jamaicans and the desacralizing of the beingness associated with such bilingualism. Relocating the people's language within sacred spaces is a necessary substantiation of creolization, the transformative force of the collective and individual psyche of postcolonial human being(s). And transformation is the product and vehicle of that psyche's artistry.

How then can we sing when our language is stigmatized and desacralized? We sing in and from creolization. Some countries have four seasons in a year. In the Caribbean, we have many. The seasons include mango, breadfruit, tamarind, ackee, guinep, and crop over, among others. Then there are the rainy seasons, the hurricane seasons, and the dry seasons. Together, these are fruits of creolization and are the temporal markers of a year. In our informal and formal international relationships, spring, summer, autumn, and winter replace these markers. Thus, we highlight the preclusion of Creole languages as respectable language forms and as languages of production.[15] Glissant (1996, 187) opines, "Creole is impoverished because terms relating to professions disappear, because vegetable oils disappear, because animal species disappear, because a whole series of expressions that were linked to forms of collective responsibility in the country are disappearing as this responsibility diminishes." Glissant proffers that the dominance of formal French is connected to language's ability to give expression to technology and machinery. Hence, a language can

---

14. The use of "bilingual" does not indicate fluency in producing two different languages. It is a more nuanced idea that recognizes a spectrum of linguistic capacities ranging from fluency in language production to the ability to understand two languages. Yet see Warner-Lewis (1996) for an investigation and analysis of the retention, influence, and social significance of Yoruba in Trinidad.

15. Jamaican, the language of Jamaica is no longer a Creole language. It is a signifier of creolization and therefore valuable and useful in ways that disavow Glissant's disregard for language as Creole.

be "liturgicalized" in its connection to domination (183, 192) and, in a creolized context, desacralize the Creole tongue and its categories.

Creolized singing is a verbalization of Caribbean sacredness. Calypso, Soca, and Reggae are the popular musical traditions of the Caribbean. Popular music has practiced, in form and lyrics, a Caribbean version of disestablishmentarianism. It is the music of the unfettered soul.[16] Singing and musicality connect to Glissant's turn to artistic aesthetics in poetry, plays, and novels as *oraliture*[17] as the modality of aesthetics most attuned to creolization and Creole. Lucy Bailey was inside her skin when she spoke privately to God. Because of lifelong devoutness, the liturgical affirmation of British English was her snare. That snare is a significant contributor to desacralization. Unfettered musicality and *oraliture* point to the language and creativity of Caribbean persons' and cultures' artistry. These aesthetic manifestations communicate with the deep psyche and restore that which is lost through cultural prejudice. The mothering tongue is "indispensable in all cases to psychological, intellectual, and emotional equilibrium among members of a community" (Glissant 1996, 183). Artistry resuscitates the mothering tongue. Within the community, emotional equilibrium, political will, and cultural authenticity are indexes of a restored sacredness.

The Lucy Bailey anecdote indicates the importance of analyzing the impact of colonial liturgical language. In the exploration of Caribbean hermeneutics, it seems vital to develop a full understanding of the sociocultural impact that our linguistic choices enforce. The proposition that the approbation of linguistic stigma through liturgy reinforces desacralization deserves to be analyzed in order to open space for biblical hermeneutics that are as creative and generative as the other sacred space in the Caribbean, "secular" culture. When one dispenses with social compartmentalizations that separate the sacred and the culture, then it is possible to perceive the ways in which sacred signifiers emerge from within cultural space. Sacred symbols, narratives, and chants in their religious tonalities share common

---

16. By unfettered soul, I intend to connect the mentioned music forms to artists who are disconnected in creative and generative places from the restrictive anglicized musicality of historical Protestantism and European styled Roman Catholicism in the Caribbean.

17. *Oraliture* is a Haitian neologism. Glissant (1996, 188) sees the invention of that term to be an indication that orality is not passé. To Glissant's list, I add art, sculpture, handicrafts, and storytelling.

fabric with music, dance, folktales, riddles, proverbs, crafts, food, poetry, and the senses that articulate divine productivity in primordial and quotidian times. The spheres of life and events are sites of encounter where, for example, "ancestral spirits return into the human world to set right human affairs that might, by disorder, incur the wrath of the gods" (Kalu 2000, 61). Dominique Zahan makes a more comprehensive statement. "For [Africans] everything that surrounds them exhibits a sort of transparency that allows them to communicate directly with heaven. Things and beings are not obstacles to the knowledge of God; rather they constitute signifiers and indices which reveal the divine being" (Zahan 2000, 4, see also Griaule 1965, xiii–xiv). This is the most elementary kind of understanding that can be posited of culture as sacred space. For purposes of accessing vital life forces for decolonization, culture—as the recuperation of history and language and source of the human person as cultural product—is an instantiation of sacred performance. Where categories of identity formation are lost or suppressed, creolized cultural reminiscences are the sphere of recuperation: a humanistic type acknowledgment of identity recovery and soul affirmations as sacred rite in "secular" spheres. Without reversion and recuperation, our biblical hermeneutic will lack an important aspect of cultural resonance and remain detached from and resistant to the profound sacred confidence that authentic creolized culture generates. Doing a sociolinguistic analysis of the sacred/secular linguistic divide promises a performance of island disruptiveness. To not do so may be to take the "Land's name in vain."

And the Subaltern Thought While Reposing on an Island

Creolization, disruption, and orality contribute rallying points in an island-based Caribbean biblical hermeneutic. In Glissant are echoes of Homi Bhahba, Edward Said, Gyatri Spivak, and other postcolonial thinkers. The echoes suggest commonalities though the reverberations differ. Hybridity, for example, in Bhaba's work forecloses on pretensions to essentialist purity in literary representations of the colonizer. Hybridity is a counternarrative, a critique of the canon, and as Abdennebi Ben Beya (2012) put it, "the migration of yesterday's 'savages' from their peripheral spaces to the homes of their 'masters' [that] underlies a blessing invasion that, by 'Third-worlding' the center, creates 'fissures' within the very structures that sustain it." As a summary assessment of hybridity, this is sufficient. However, as echoic as Glissant might be of Bhaba, Said, and Spivak, his project and articulations differ very subtly, but importantly.

Glissant's reflections and critique are born out of agony over the incapacitations and subjugations of his Martiniqan compatriots and of hope undeterred. The basis of Glissant's theorizing is the creolized person on a landmass that should have been her or his own. Because the Martiniqan situation is drastic, it presents a hyperbole of the general situation in the Caribbean where islanders' struggle for identity is a process of overcoming and surpassing the point of entanglement. Bhaba, Said, and Spivak analyze the self-aggrandizing literature of the masters as they represent themselves to their own and represent the other with exoticizing tendencies and an entrenched orientalism. Glissant, *aull contrairy*,[18] reflects on another originary situation. His is the lived reality of the people who made home away from the home of dispossession, disconnection, and dislocation. Glissant discusses their aesthetic productivity as a sign, expression, and mechanism of hope. What is the function and potential efficacy of ethnocultural musicality and *oraliture*? It disrupts Eurocentric History, and any totalizing tendencies that appear in the Western co-optation of postcolonial metanarratives. Glissant, I daresay, might see in James neither mimicry nor hybridity but resistant, subversive creolization[19] and the arts as vehicles of transformation.

The difference between Glissant's rupture in history and Bhaba's gaps in literature needs resolution (Williams and Chrisman 1994, 76). In Glissant, both history and literature, done by the colonizer, ideologize "the spread of a dominant sameness" (see Glissant 1996, 69–70 for a fuller discussion). Thus, Bhaba's gap hints at the rupture in the collective memory and culture of the colonized of which Glissant is aware. The first rupture was the slave trade, the second rupture, is posttranslantic history. History has to enable peoples to connect "the accumulation of [their] experiences (what we would call its culture)" as part of a continuum with a global historical consciousness. Without that connection, "history as far

---

18. I am here practicing what I might designate as a "Glissantism." I am creolizing the French term *au contraire* in a way that deconstructs through punning while producing and introducing the intentionality of contrariness that marks Glissant's work and characterizes his response to nonisland, non-Caribbean postcolonialism.

19. Glissant (1996, 89–91) gives a fuller discussion of the periodization of Martiniqan history. Glissant's and James's strategies similarly place a marginalized people at the center of history. In so doing, Glissant disconnects Martiniqan history from French history. Instead of categories such as the "Third Republic" or the "Interwar Period," for example, Glissant uses, the "Slave Trade/Settlement," the "Plantation System," "Legitimized-Legitimizing Assimilation," and the "Threat of Oblivion," for example.

as it is a discipline, and claims to clarify the reality lived by this people, will suffer from a serious epistemological deficiency" (61). The gap betrays the ideological implausibility of colonial literature. The rupture is the disconnection of peoples from their ancestral past, the immediate past of their present. In Glissant, reversion is the strategy that dismembers gaps and ruptures, shreds veils, and re-members history disruptively.

Concluding Foreword

Ultimately, Glissant's response to his own analysis lies not in triumphalism or in nativist naïveté. Rather, he perceives that civilization is undergoing a massive transformation that heralds a shift in global historical and literary relationships (Glissant 1996, 97). Two elements in the transformation are diversity that represents the human spirit's striving for a nontranscendental crosscultural relationship and a movement from the written to the oral. Diversity, to be successful has to be a crosscultural relationship devoid of transcendence (98; see also Glissant 1997 and Miguel 2009), that is, without pretensions to timelessness and unlimited spatiality. Writing will be salvaged by orality. Glissant (1996, 101) advises, "If writing does not henceforth resist the temptation to transcendence, by, for instance, learning from oral practice and fashioning a theory from the latter if necessary, I think it will disappear as a cultural imperative from future societies." His is a vision in which the oral and the emergence of "oraliture" (245) as creolized writing, by the poets, the artists, and the singing people correct the written. Glissant's confidence in Creole as an oral source of creativity points to a radical *oraliture* as the greenhouse of hermeneutics. And ceaseless creolizing, as a dynamic signifying upon the past, the present, and future productivity, can be the playground of Caribbean hermeneutics. This would be rupture disrupted and repaired in the real, where people—Martiniqans, Caribbean islanders—live resisting the vile veil. But more, it puts the empire on alert: ceaseless creolization is contagious (see Bernabé, Chamoiseau, and Confiant 2005, 284).

Yes, as the (eye)land plays with itself, the sea, the horizon, and a vast beyond, should you pursue the pull, another (eye)land appears. How playful, how curious, how Creole, how (eye) is land. The drum beats a ready rhythm, the (eye)land *transmoots* its undulating *seaing, w(e)aving*. Let's Creole!

Works Cited

Barry, Peter. 2009. *Beginning Theory: An Introduction to Literary and Cultural Theory*. 3rd ed. Manchester: Manchester University Press.

Bayoumi, Moustafa, and Andrew Rubin, eds. 2000. *The Edward Said Reader*. New York: Vintage House.

Ben Beya, Abdennebi. 2012. "Mimicry, Ambivalence, and Hybridity." Postcolonial Studies @ Emory. https://scholarblogs.emory.edu/postcolonialstudies/2014/06/21/mimicry-ambivalence-and-hybridity/

Bernabé, Jean, Patrick Chamoiseau, and Raphaël Confiant. 2005. "In Praise of Creoleness." Pages 274–89 in *Postcolonialisms: An Anthology of Cultural Theory and Criticism*. Edited by Gaurav Desai and Supriya Nair. New Brunswick, NJ: Rutgers University Press.

Bhabha, Homi. 2005. "Of Mimicry and Man: The Ambivalence of Colonial Discourse." Pages 265–73 in *Postcolonialisms: An Anthology of Cultural Theory and Criticism*. Edited by Gaurav Desai and Supriya Nair. New Brunswick, NJ: Rutgers University Press.

———. 2008. *The Location of Culture*. New York: Routledge.

Césaire, Aimé. 2000. *Discourse on Colonialism*. With introduction by Robin D. G. Kelley. Translated by Joan Pinkham. New York: Monthly Review.

Dash, Michael. 1996. Introduction to *Caribbean Discourse: Selected Essays*, by Édouard Glissant, trans. Michael Dash. Charlottesville: University Press of Virginia.

Desai, Gaurav, and Supriya Nair, eds. 2005. *Postcolonialisms: An Anthology of Cultural Theory and Criticism*. New Brunswick, NJ: Rutgers University Press.

Dubois, W. E. B. 1903. *The Souls of Black Folk*. New York: Cosimo Classics.

Edmondson, Belinda, ed. 1999. *Caribbean Romances: The Politics of Regional Representation*. Charlottesville: University Press of Virginia.

Glissant, Édouard. 1996. *Caribbean Discourse: Selected Essays*. Charlottesville: University Press of Virginia.

———. 1997. *Poetics of Relation*. Translated by Betsy Wing. Ann Arbor: University of Michigan Press.

———. 2001. *The Fourth Century: A Novel*. Translated by Betsy Wing. Lincoln: University of Nebraska Press.

Griaule, Marcel. 1975. *Conversations with Ogotemmêli: An Introduction to Dogon Religious Ideas*. London: Oxford University Press.

Iyasere, Solomon O. 1976. "Cultural Formalism and the Criticism of Modern African Literature." *The Journal of Modern African Studies* 14:322–30.

James, Leslie. 2010. "The African Diaspora *as* Construct *and* Lived Experience." Pages 11–18 in *The Africana Bible: Reading Israel's Scripture from Africa and the African Diaspora*. Edited by Hugh R. Paige Jr. Minneapolis: Fortress.

McFarlane, Adrian, 1998. "The Epistemological Significance of 'I-an-I' as a Response to Quashie and Anancyism in Jamaican Culture." Pages 107–21 in *Chanting Down Babylon: The Rastafari Reader*. Edited by Nathaniel Samuel Murrell, William David Spencer, and Adrian Anthony McFarlane. Philadelphia: Temple University Press.

Míguez, Néstor, Joerg Rieger, and Jung Mo Sung. 2009. *Beyond the Spirit of Empire: Theology and Politics in a New Key*. London: SCM.

Nardal, Paulette. 2009. *Beyond Negritude: Essays from "Woman in the City*. Translated with an introduction and notes by T. Denean-Sharpley-Whiting. Albany, NY: State University of New York Press.

Nesbitt, Nick. 2003. *Voicing Memory: History and Subjectivity in French Caribbean Literature*. Charlottesville: University of Virginia Press.

Said, Edward. 2002. *Reflections on Exile and Other Essays*. Cambridge: Harvard University Press.

Senghor, Léopold Sédar. 2005. "Negritude: A Humanism of the Twentieth Century." Pages 183–90 in *Postcolonialisms: An Anthology of Cultural Theory and Criticism*. Edited by Gaurav Desai and Supriya Nair. New Brunswick, NJ: Rutgers University Press.

Sommer, Doris. 2000. "A Vindication of Double Consciousness." Pages 165–79 in *A Companion to Postcolonial Studies*. Edited by Henry Schwarz and Sangeeta Ray. Malden, MA: Blackwell.

Sugirtharajah, R. S. 2012. *Exploring Postcolonial Biblical Criticism: History, Method, Practice*. Malden, MA: Wiley-Blackwell.

Walcott, Derek. 2005. "The Caribbean: Culture or Mimicry." Pages 257–64 in *Postcolonialisms: An Anthology of Cultural Theory and Criticism*. Edited by Gaurav Desai and Supriya Nair. New Brunswick, NJ: Rutgers University Press.

Warner-Lewis, Maureen. 1996. *Trinidad Yoruba: From Mother Tongue to Memory*. Tuscaloosa: University of Alabama Press.

Williams, Patrick, and Laura Chrisman, eds. 1994. *Colonial Discourse and Postcolonial Theory: A Reader*. New York: Columbia University Press.

Zahan, Dominique. 2000. "Some Reflections on African Spirituality." Pages 3–25 in *African Spirituality: Forms, Meanings and Expressions*. Edited by Jacob K. Olupona. New York: Crossroad.

Gaelic Psalmody and a Theology of Place in the Western Isles of Scotland

Grant Macaskill

The Western Isles of Scotland, the Outer Hebrides, located in the far northwest of the country, now constitute, along with Skye, the last part of the country in which Gaelic language and culture are maintained in any extensive way. The survival of Gaelic in these areas is due largely to the insular[1] geography and to the ability of island communities to maintain cultures lost in mainland contexts. It is also due to the presence on those islands of strong religious communities for which the Gaelic language is a major part of cultural identity. Those communities vary from island to island: some islands are strongly Protestant, Calvinist, and Presbyterian, while others are strongly Catholic. For all of these groups, Gaelic language is an important part of their cultural and spiritual identity. In the case of the Presbyterian churches, though, the practice of exclusive and unaccompanied Gaelic psalmody is a distinctive element within that identity, one so central that the thought of abandoning it would be devastating to the subculture.[2] In such worship, sections of psalms are sung unaccompanied, and they are sung exclusively; there are no hymns, no

---

1. My use of this adjective throughout this chapter reflects its technical sense: pertaining to islands/an island.

2. In November 2010, just three days before I presented this paper at the Society of Biblical Literature, the Free Church of Scotland, the major Presbyterian denomination in the Western Isles, voted to permit congregations to adopt practices other than exclusive psalmody. What is interesting about this decision is that the votes of representatives of the Western Isles Presbytery were crucial to securing the permission. Those representatives, influenced by an erudite paper on the biblical evidence by Rev. Alasdair I. Macleod and by eloquent and persuasive speech by Rev. Dr. Iain D. Campbell, deemed the unity of the church to be more important than the maintenance of exclusive psalmody. The Western Isles Presbytery, then, was of huge importance in securing

choruses, and no instruments. While the practice of exclusive psalmody is not unique to these communities and is shared with the mainland denominations to which they belong (historically it was the practice of the Scottish church), it has never had quite the same role in shaping identity outside of the *Gàidhealtachd*, the Gaelic speaking world of the Highlands and islands. The musical form of specifically Gaelic psalmody is highly distinctive, particularly in the context of western European music, and is almost impossible to either imitate or even to participate in for those who have not grown up in Gaelic churches.

I offer, in this essay, that the centrality of Gaelic psalmody to the identity of island Christians, living in rural communities, participates in a strong theology of place: a profound awareness that God cares for and engages with specific places. This theology may not be explicitly articulated, but emerges in the attitudes to the island and its nature. While a distinctively insular feature of island Christianity, I further suggest that reflection on this fact may help us to see how the use of the psalms in worship outside of the island context may inform a more robust theology of the environment, with implications for how we evaluate specific places and their use, though always with less thoroughgoing effect than the psalms will have upon the islander. My development of this argument reflects my own status as an insider within the island church; in the absence of adequate documentation, I draw upon my own experience and tradition and ask the reader to take this on trust. At the same time, I support my claims with further lines of evidence that will be more accessible to the reader, although this will involve some more tangential argumentation.

1. The Western Isles: Geography and Culture

Before we proceed to an examination of the relationship between exclusive psalmody and place, we need to have some sense of the place(s) in question, so a brief description of the Western Isles is necessary. Various names are applied to this archipelago, and these may be encountered in the historical literature on them, as well as in the maps of western Europe. The Romans named the islands off the Scottish west coast the Ebudae, which became Hebudes and eventually Hebrides. A traditional differentiation is

---

the permission to sing other material; to this day, however, no island congregation has considered adopting any practice other than exclusive, unaccompanied psalmody.

made between the more northern Outer Hebrides (Lewis and Harris, Benbecula, Barra and the Uists, Saint Kilda and the various tiny islands that exist as satellites off their coasts) and the more southern Inner Hebrides (Skye, Mull, Iona, and further clusters of tiny islands). That distinction is reflected in one of the sets of Gaelic terms applied to the islands: the Outer Hebrides are *na h-Eileanen a-Muigh*, and the Inner Hebrides are *na h-Eileanan a-Staigh* (the outer and inner isles, respectively). The broadly synonymous Gaelic terms for the Western Isles, *Na h-Eileanan Siar* and *Na h-Eileanan an Iar* tend to be confined to the Outer Hebrides. One further term is often encountered, and it is enlightening: *Innse na Gall*, the Islands of the Foreigners. The label calls attention to the Viking past of these islands and to the hybrid identity of their people, a blend of Celtic and Norse culture and genes, the latter seen most clearly on Lewis, the northernmost of the islands, where the place names are Norse in origin and where the spoken Gaelic is of a quite different tonal character to the other islands. Recent genetic studies have confirmed the continued prominence of Scandinavian gene-types among islanders. History and tradition also attest the lingering presence of Scandinavian paganism among the people of the islands, even up to recent times (see MacLeod 2008, 96–99).[3] Once widespread in the Highlands, Gaelic is now spoken to any extent only in the Outer Hebrides and in Skye, having died out almost entirely on the mainland and on the other islands, which suffered more severely from depopulation of indigenous peoples.

Although some of the islands are more fertile than others, with Barra and the Uists in particular blessed by more extensive tracts of *machair*—well-drained sandy coastal grassland fertilized by seaweed, full of meadow flowers, and capable of being farmed—most of the land of the Outer Hebrides is wind-scoured and barren, with almost no tree life, a thin and acidic peat topsoil capable of growing only heather, and an impermeable rock beneath the soil; combined with the heavy island rainfall, borne in on regular fierce Atlantic storms, this means that the land is peppered with small lochs and that where the topsoil is of any thickness it turns to bog.

Strips of peat are cut every year, in the spring, and left to dry over the summer, serving as the major fuel source for rural islanders during the

---

3. MacLeod's book, while not an academic study, is the most serious and thoroughgoing study of the religious history of Lewis and Harris (confusingly, both "islands" belong to the same land mass, often referred to as the "Long Island," though culturally they are quite distinct and to confuse the two is offensive).

winter months. The activity is a communal one, with neighbors helping one another to cut and stack the peats as well as bringing them home. The scale of this harvest makes little impact on the peat stocks on the island, since the cutting is done by hand. There is a strong sense of connection, then, to the soil of the island that will warm the island house; the smell of burning peat continues to be one of the strongest associations that an exiled (see below) islander will have with home.

The nature of the soil also dictates the kind of agriculture traditionally practiced in the islands: a subsistence farming known as crofting involves the keeping of small numbers of sheep, usually the small Scottish Blackface, potato and carrot farming, and where possible (and this is by no means everywhere) the growing of small amounts of arable crop for grains and hay. With nearly every island village being located on the coast, fishing for herring and mackerel supplements this, though such fish are a seasonal harvest, meaning that salt fish is one of the traditional components of the island diet, though its place has diminished as the island economy is modernized. Larger scale fishing, particularly trawling, has also been an important industry in the islands. Crofting is now in serious decline, partly through the depopulation of the islands and the incoming of "settlers" from the mainland, partly through the difficulty of sustaining such an industry in the face of modernity.

The service sector has grown in recent years, as faster and cheaper ferries[4] have meant that tourism to the islands has increased, and beyond that much of the employment is of the kind associated with supporting a community of this size. Although there has been some resurgence of the Harris Tweed industry in the islands, generally the islands have a fairly low employment rate and a growing population of retired people, particularly incomers.

The islands have a rich, if neglected history. As well as being home to some of the earliest archaeological evidence in Britain—the standing stones at Callanish (*Calanais*) being dated to at least 2000 BCE, if not earlier—the islands have some of the least spoiled remains of the Celtic church, with the memory of the Columban order still echoed in some of the place names, and significant witnesses to the period of Norse rule, most sensationally the Lewis Chessmen found at Uig and also a number

---

4. The crossing from Ullapool to Stornoway now takes only two hours and forty-five minutes. In my childhood, it would generally be closer to four hours.

of Norse mills.⁵ Some of the customs of the islands are also seen as living witness to the Norse period. The history of the islands, caught up in the Viking conflicts and in later Scottish politics and clan wars, involved a good deal of bloodshed, but the worst of this was actually seen in relatively recent times. In the nineteenth century, under a regime of landlords who had sought to limit crofting in order to force islanders to work in the brutal industry of manufacturing lime from kelp burning, starving islanders, little better than slaves, raided deer parks and engaged in uprisings in Lewis, Harris, and Skye, a bloody action that eventually forced the Westminster Government to defend their rights in the Crofters Act (see Buchanan 1996).

Despite this act, these land raids were mirrored by similar events in the early twentieth century, as crofters rose up against Lord Leverhulme, who had sought to industrialize the island, denying crofters access to land that they had been promised in connection with their participation in the wars and creating serious poverty among the rural communities. This last point also brings us to one final striking feature of island culture: historically, the men of the islands were celebrated mariners, caught up in war efforts, serving in the navy and in the merchant navy, as well as in the industry of fishing. The sea has taken more island lives that can be counted and all families remember those that they have lost to it, but it has also borne islanders around the world and, in many cases, home again.

Religion continues to be a significant factor in the islands. The southern islands of the Outer Hebrides (Barra, South Uist, and, to an extent, Benbecula) are strongly Roman Catholic, while the northern islands and Skye are strongly Presbyterian. Within the latter, the greatest proportion belong to the Free Church of Scotland and to its sibling, the Free Presbyterian Church.⁶ These churches trace their lineage to the Reformation in Scotland, through the Disruption of 1843, when a large group of evangelicals broke from the Church of Scotland over the issue of patronage, the right of a landlord to sponsor and choose a minister for the local congregation, and elected in the process to become disestablished from the state. Outside of the islands, these denominations are small, though with

---

5. MacLeod (2008, 49–70) provides an excellent discussion of the Celtic and Norse history and legacy.

6. An ethnographic study of village life in Lewis was published by Susan Parman (2005). The seventh and eighth chapters provide discussion of religious practices in the islands. See also MacLeod 2008, 279–342.

pockets in the major cities where they are stronger, often by displacement of islanders. Both churches have maintained the practice of exclusive psalmody, to the extent that it is closely woven with their identity, as well as a strong Sabbatarianism, which has ensured that Sunday ferry services to the islands have been prevented until recently. Services are still conducted in Gaelic, though English services are now also held. It is worth noting that Gaelic is much stronger in the villages than in the one major island town of Stornoway, but even in the villages it is on the decline, as depopulation and immigration takes effect.

This lengthy description of the islands is quite important in impressing upon the reader the cultural identity of islanders. All islanders have a strong sense of belonging to a place that the outsider would regard as bleak and barren; the number of islanders who return to the islands after studying elsewhere is quite striking, and all the more so when the limited employment offered is recognized. Those who live away from the islands will often consider themselves to be exiles (Parman 2005, 66–69, 155–56), seeking out others from their island in bars[7] and churches.[8] Whether or not they still practice crofting, most outside of Stornoway—the one large town on the islands—have a strong sense of the land and of its importance to their own lives, even if it is just in terms of the peat that continues to be burned. Whether in terms of their own experience or in terms of their parents', they will also have the sense of dependence upon divine providence in nature that only a subsistence farmer can have. In the villages, native islanders also have a strong sense of the decline of Gaelic and of Gaelic culture and of the death of ways of life for which, in many cases, their great-grandparents suffered and fought. It is in such terms that the islandedness of the Western Isles has to be considered.

2. The Form and Practice of Gaelic Psalmody

Gaelic psalmody is an unusual and distinct musical form in the Western tradition, although it has recently been argued that it has left a much greater legacy than most are aware of through a formative influence on black gospel music and Native American church music (for the case and some criticisms, see Miller 2009, 243–59). This case has been challenged

---

7. The Park Bar in Glasgow was a renowned meeting place for Lewis exiles.

8. In my own time in Glasgow, Partick Highland Free Church was explicitly associated with the Lewis exiles and Grant Street Free Church with Skye exiles.

and should certainly be treated with caution, but it has opened up fruitful cultural engagement between black churches and choirs and Gaelic ones.[9]

The psalms are sung unaccompanied, with a precentor lining out the opening phrase of the verse and then the congregation joining in to sing the rest of the line (see Miller 1984, 15–22). There is no defined rhythm and singers move from note to note in their own time and by their own route, gliding between notes rather than jumping to the next note. Grace notes, of the kind also found in the piping tradition, are commonly used, with studies suggesting that single notes will often be adorned by fourteen grace notes.[10] The result is a typically slow, freeform music with unique and spontaneous harmonies that do not accord with formal harmonization, transient discordances that resolve naturally and a swelling and ebbing quality as voices merge and part.[11]

Over the years, I have typically heard Gaelic psalmody described in two seemingly contradictory ways. Often it is described as an "alien" sound, given that it is so unlike anything else in Western tradition. There is certainly a somewhat otherworldly quality to the novice. By contrast, it is also often described as "a force of nature," reminding hearers of the sound of the wind or of the sea, two sounds any Western Islander is familiar with. I am not competent to do it, but it might be interesting to examine whether the natural world experienced in the *Gàidhealtachd* has influenced the form of musical expression.

3. Exclusive Psalmody and a Theology of Place

Psalms and psalms alone (where psalm is narrowly defined as songs from the biblical Psalter) are sung, and they are always sung unaccompanied. This is interesting because it means that Christians are forced, consciously or subconsciously, to translate what they are singing to be *their* song of worship. This obviously involves a Christological aspect: there is a strong tradition of seeing Christ in the psalms as the object, and sometimes the

---

9. This included a major collaborative performance at the Celtic Connections festival in Glasgow, 2005.

10. The study is noted by MacLeod (2008, 289), but with no bibliographic detail. I have been unable to locate the study.

11. High-quality audio samples can be accessed at "Listening Lounge," 2007.

subject,[12] of worship. This is not unique to Gaelic Presbyterianism, of course, but it is much more thoroughgoing in that tradition.

More subtly, the translation process also involves geographical translation. Many of the psalms root their description of the believer's experience of God in specific times and places that are foreign to contemporary island worshippers, in the ancient Mediterranean and Middle Eastern worlds. When these places are such a prominent part of worship, singers must either appreciate the symbolic value of the place spoken of (as, for example, in the case of Zion) or they must see the psalm's place of encounter as a parallel to their own place of encounter. While to sing of Zion and Jerusalem may be to draw upon place names with symbolic value and association, to sing of "Baca's vale" as a place of refreshment (Ps 84) or the Negev as a place of transformation (Ps 126) requires a sensitivity to the encounter of the believer with divine grace in specific ways in specific places. It requires a sense that these places are part of his economy, that encounter with God is something that takes place within the circles of this world, in particular places at particular times in particular ways. To sing such verses in the Western Isles involves a geographical translation: the wilderness of Judah (Ps 63 and other references to "wilderness") or the desert of Kadesh (Ps 29) become Barvas moor, flooded with autumn rains; the heights of the hills (Ps 18.23, Ps 121:1) become the mountains of Cliseam, Crogearraidh, or Sgurr Alasdair. The act of singing becomes an exercise of the imagination, mapping psalmic language onto island space, as an articulation of the placed encounter with God.

But in addition to the emphasis that the believer's encounters with God are earthed in specific places, there is also a sense that those places themselves participate in God's gracious dealings with the world. The psalms' utilization of the natural world as revelatory of God's power and, as importantly, his providential care for creation (Ps 29, Ps 103) are also translated, and here the distinctively rural character of the island communities plays a crucial interpretative role. The relationship with the natural world is unavoidably immanent in communities based around smallholding crofts, particularly where the act of farming is a difficult one, with limited returns from a ground of limited fertility. The act of trusting in the care of God for God's creation is a living one in such a context; an

---

12. Notably, this is the case with Ps 22, understood in toto to be the song of Christ upon the cross.

experience that the urban dwellers of the world, including those in the large island town of Stornoway, have been dislocated from. I do not mean by that that they do not trust God and do not have to exercise radical faith, but rather that faith is not necessarily focused on God's providential lordship over weather and tide, et cetera.

The component parts of the Hebridean world, though, are quite different from those of the Levant; there are virtually no trees in Lewis or Uist, for example, so how do we picture the breaking of the cedars of Lebanon by the voice of God (Ps 29:5)? Again, then, the worshiper is required to translate the image into his or her own experience, finding local analogues for the imagery. There is no question, though, that this act of translation retains the insight of the agrarian into the engagement of God with the natural world and that this reflection is focused on his engagement with specific places. By contrast with the inward piety reflected in many contemporary worship songs—Jesus and me: a different kind of insularity!—the practice of exclusive psalmody in such a rural context is outward, forcing by its exclusivity the recognition that the believer's encounter with God is *placed* within a wider relationship between creator and creation. The place of the island in the divine economy is thus affirmed.

It is important that we recognize that this theology of place is not generated by the psalms in themselves or even by the practice of exclusive psalmody, but rather by the symbiosis of practice and place. Outside of the islands, with their preservation of Gaelic culture in a rural context, exclusive psalmody does not necessarily have such force, and perhaps this is why it has had less centrality to Christian identity elsewhere. Within the insular context, however, the combination of psalmic language of place and islanded identity is a potent one, creating a strong sense of the place in God's dealings with the world. Scratch the skin of any island Christian, and you will find a sense that their spirituality, their faith, is grounded in the island. Away from the island, as many ultimately are, they are bereft of something important, not simply their home church, but something more basic, more elemental: their place.

There is a specific Gaelic word for this sense of loss: *cianalas*. This word denotes the essential yearning for home, where home is place, rather than the more sentimental or social focus of the English "homesickness." This longing for a place is married in interesting ways to faith in island Christianity, with the island given a significance that is almost akin to the temple. Recent debates about Sunday sailings reflect this: for Christians in the islands, the issue is not simply about defending a way of life, it is about protecting the sanctity of the place (whether we agree with them or not). That is why their protests are so loud. That is also why most of the voices that support change most aggressively are those of the incomer.

There is a danger of lapsing into a naïve romanticism here, though, and we must be careful. This sense of the island's significance does not necessarily correspond to a healthy ecotheology. It is certainly true that one will find a lack of respect or love for the environment in the Western Isles just as we might find it elsewhere. A recent debate took place in the islands over whether to construct the largest windfarm in Europe on the huge moorland area of Barvas. Those who advocated such a development did so on largely utilitarian grounds: what good is this moor for anything else? That utilitarianism, though, failed to win the argument and led to major schisms within the island and nation: naturalists were concerned at the impact on one of the last major areas of bogland in Europe; ornithologists raised concerns about birds. For most of the islanders who objected, however, the concern was simply that this *place*, this environ, would be changed beyond recognition. The debate highlighted the extent to which many islanders regard the place as somewhere to be treated with a measure of reverence.

4. Confirming the Influence of the Psalms

It is possible that what we are describing here is simply a matter of coincidence: that islanders who love their islands also happen to sing the psalms exclusively. One of the difficulties faced in the task of verifying the kind of thesis I am advancing is that much of the evidence that would support the influence of the psalms is in the form of oral testimony and conversation. Believers will describe their conversions or their ongoing experiences of God's typically using the language of the psalms, and often directly quoting those psalms, but such testimonies have not been well documented. What is needed is a serious attempt to examine and contrast the spiritual vocabularies and attitudes of islanders from the psalm-singing tradition with those from other, for example, Baptist or Catholic traditions. Again, the extent to which insularity allows the preservation of distinct traditions may facilitate such a study.[13]

---

13. A good deal of ethnographic work, such as that of Parman (2005), has been done in the islands, but this seems to have been a blind spot in the research.

John MacLeod's study of island religion does record some examples of the use of psalms that support my claim, though. The singing of Ps 122 by Christians crossing from Carloway to Bernera for communion is one such:

> Pray that Jerusalem may have peace and felicity,
> Let them that love thee and thy peace have still prosperity. (MacLeod 2008, 275–76)[14]

This example warrants a little more elaboration. In the Presbyterian culture of the islands, communion takes place only twice a year, within the context of a "communion season." Such seasons involve an extended series of meetings, both in the churches and in the homes of believers, for preparation and then thanksgiving for the communion itself. Local congregations schedule their meetings to allow members to visit other communion season meetings around the island. The churches maintain a tradition of singing certain psalms around this time, notably sections of the Hallel Psalms (113–118) and the Psalms of Ascents (120–134), consciously mirroring the practice of pilgrims visiting Jerusalem for Passover at the time of Jesus. The singing of Ps 122 in the example above is part of this tradition and ascribes to Bernera, where the communion takes place, the status of Jerusalem. It is the place where the presence of God will be experienced in a special way by God's people. Although the particular location of communion might change, and hence other villages would take the place of Jerusalem, the practice constitutes a distinctive reactualization of the meaning of the psalm, affirming the island as the place invested with the glory of God, in the context of a local emulation of the practice of pilgrimage.

A rather darker example is found in the use of the Gaelic of Ps 77 on the monument to the Iolaire tragedy, in which hundreds of island men, returning home after the First World War, were drowned within sight of their homes in Stornoway harbor.

> *Do cheuma tha 'san doimhneachd mhoir, do schliche tha's a'chuan:*
> *ach luirc do chos cha-n aithnich sinn, tha sud am folach uainn*

> Thy way is in the sea, and in the waters great thy path.
> Yet are thy footsteps hid, O Lord; none knowledge thereof have. (See MacLeod 2008, 234).

---

14. Here, and below, I will provide only the English translation.

In representing the mysterious sovereignty of God over this act of providence, it is specifically a psalm that is chosen for the memorial. The key point is that this regular item of island worship, Ps 77, is understood to affirm God's sovereignty over the seas around Stornoway harbor in which so many had died. The quotation from the psalm responds to doubts over his sovereignty that might arise from the devastating events, since it affirms alongside his sovereignty the inscrutability of his will.

These two examples highlight, in quite different ways, the extent to which the exclusive use of the psalms in worship informs a perception of God's direct involvement with this specific place. In fact, it is interesting that the dedication page of MacLeod's spiritual history of Lewis and Harris includes a quote, in Gaelic, from Ps 36:6, 7.

> Your righteousness is like the mighty mountains,
> your judgments are like the great deep;
> you save humans and animals alike, O LORD.
> How precious is your steadfast love, O God!
> All people may take refuge in the shadow of your wings. (NRSV)

This takes us some way toward confirming my proposal, though the evidence still falls short of proof. In part, this shortfall reflects the lack of thorough and systematic documentation of religious practice in the islands, in terms of both oral accounts of island practice (such as communion) and records of inscriptional evidence, such as that found on memorial stones and gravestones. Advances in these areas would allow the examples that I have cited to be contextualized more fully and examined more systematically.

In the absence of such documentation, though, we may draw upon a more creative approach to add further support to the thesis. Elsewhere, I have explored the influence of the theology of place in the psalms on the work of the island songwriters Calum and Rory MacDonald.[15] The brothers, principal songwriters for Scottish rock band Runrig, have over many years supported the study and promotion of Gaelic psalmody, their record label sponsoring the first professional recordings of congregational Gaelic singing. There are two interesting examples of the influence of the psalms

---

15. I delivered a paper on their work at the "Reversed Thunder: The Art of the Psalms" conference of the Institute for Theology, Imagination, and the Arts, University of St Andrews, 2009.

on their songwriting that support the conviction that psalmody has fostered a theology of place.

The first is in the song "This Darkest Winter" (1985), which describes an experience of conversion, as the "blinding lines" of Scripture drive away the shadows from the writer's soul.

> And then I saw a distant sight, a heart behind the grey,
> Come shining through the darkest night, establishing my way. (MacDonald and Macdonald 2001, 274–75)

Those closing words are lifted from the old Scottish Psalter translation of Ps 40:2, and the allusion is explicit to anyone familiar with that volume, as most in the Gaelic psalmody tradition would be. Strikingly, the experience described takes place "on a long dark loch, on a Uist moor." Laden with metaphor, the song does not necessarily describe a particular event that took place upon this moor, but rather captures the experience of personal spiritual transformation using the imagery of the island and fusing it with the words of Ps 40. Nevertheless, the significance of the place as the informing context of the writer's sense of restoration is clear: as the psalmist is lifted from the miry pit and set upon a rock (Ps 40:2), so the songwriter is led across the bogland of the Uist moor, "over last year's rotting corn," to spiritual renewal. A second example is less textual and more formal in character. The brothers (and Runrig) have drawn on Gaelic psalmody as a musical form in some of their own compositions, explicitly presenting these songs as inheriting a tradition of psalmody. The clearest example of this is the song "An Ubhal as Airde" (The Highest Apple) (1988), which uses the musical form of unaccompanied psalmody in its repeated chorus, blending recordings of church singing with the band's performance of their own words, in order to make the song sound like a congregationally performed psalm. It is a striking feature of this song that the musical form, deliberately modelled on psalmody, is partnered to lyrics that reflect upon how the songwriter has encountered God in the islands.

> Comhla rium
> A tha thu an drasd
> Mo shuilean duinte, mo chuimhne dan
> Nam sheasamh a' coimhead
> Gach cnoc is gach traigh
> Is an siol a dh'fhag thu ann a 'fas

Tha an garradh lan
Le craobhan treun
Le meas a' fas dhuinn ann ri bhuain
Ubhlan abaich
Milis, geur
Ach tha aon ubhal nach ruig sinn idir air

Is co 'nar measg
A mhaireas la
Seachad air am is air oidhche fhein
A liuthad uair
A shreap mi suas
Airson an ubhal as airde chur gu beul

Seididh gaoth is dearrsaidh grian
Tro mheas nan craobhan linn gu linn
Ach thig an la is thig an t-am
Airson an ubhal as airde
Air a' chraobh a bhuain

    English: The Highest Apple
You are with me now.
My eyes closed, my memory confident
Standing here watching
Each hill and shoreline
With the seed you left
Still growing

The garden is well stocked
With mighty trees
With fruit growing for the whole world
Ripe, sweet
And bitter apples
And the one apple
That is beyond reach

Who amongst us
Can exist a single day
Beyond our own time and our own limits
Countless and futile
Are times I've climbed

To reach and taste
The forbidden fruit.

The winds will blow
And the sun will shine
From generation to generation
Through the trees of the garden
But the day and the hour
Will surely come
To take the highest apple
From the knowledge tree. (MacDonald and MacDonald 2000, 50–51, reproduced by permission)

An Ubhal as Airde was written by Calum and Rory Macdonald following the death of their father. It is almost certainly their father that is addressed in the opening verse, though the psalm form of the song and the unspecified addressee leave open a certain ambiguity and the possibility that it is God who is being addressed. The pain of bereavement is probably also reflected in the second verse. This verse depicts the world as a graced reality; it is a garden, cared for and provided for, but among God's provision, depicted as fruit, are bitter apples that are unpleasant. There is much in the life that God gives us that is hard to swallow. The language here echoes the wisdom traditions of the Old Testament, not least Ecclesiastes, with its reflection on the fact that life does not always fit our believing expectations, and it is from that tradition that the song also draws its central image of perfect understanding and knowledge: Wisdom is a tree of life (Prov 3:18) and the highest apple that will one day be eaten will be taken from "the knowledge (or wisdom) tree."

While drawing upon the wisdom tradition, this image of the tree also draws upon a stock of Celtic imagery (see MacInnes 2006, 409–10). Interestingly, the same phrase (the highest apple) is used in the third verse to speak of this fruit being beyond reach. The English translation provided by the brothers renders the phrase with deliberate allusion to the Gen 2–3 tradition: it is "forbidden fruit." Perfect knowledge will be given one day, but until then faith and submission are the calling of the believer and the fruit is beyond reach, off limits.

There remains, though, a sense that God has genuinely been encountered and knowledge tasted. That encounter, moreover, is earthed in this specific place. The context for this whole faithful reflection and act of praise is the act of "standing here watching each hill and shoreline."

This line contains a subtlety lost in translation. Typically in Gaelic, the words *cnoc* (hill) and *traigh* (shore/beach) are encountered in the names of places much more than as autonomous nouns. To speak of "each hill and shoreline" in this song is to imply specific, known places, so that to catch the nuance in English, we almost have to translate the words using capital letters. The implication is that each of these specific places have played a part in the writer's relationship with his father and beyond (and perhaps through) him with God. As the chorus (the last stanza above) soars into psalmody, expressing its hope for eschatological knowledge, it is an ascent that has begun among the named hills and beaches of North Uist. The musical form of psalmody becomes a vehicle for the expression of an earthed, *placed* encounter with God.[16]

Taken together with the first two examples of the use of the psalms in the islands, these two illustrations from the music of the Macdonald brothers suggest the influence of the psalms in fostering a sense of "place" in theology. While the evidence that we have examined may not fully prove the claim that unaccompanied Gaelic psalmody has been generative of such a theology, it does support this proposal. It also suggests that further evidence might be obtained by a more systematic documentation of religious practice in the islands, including the accounts of communion practice and the study of memorial and gravestone inscriptions, and also the study of religious poetry from the islands that, like the writings of Calum and Rory Macdonald, may have been influenced by the psalms.

5. Concluding Reflections: The Psalms and the Nonislander

I have suggested in this paper that in the rural island communities, the practice of exclusive psalmody potentially furnishes a theology that takes seriously the place of *place* in God's relationship with the world. The key is that the nature of rural island communities, their relationship with the land and with the mainland, their insular preservation of Gaelic language and culture, allows for a certain fusion to take place, a mutual illumination leading to a distinct reading of both the psalms and of the place.

A proper reflection on this may help us to consider how we may nurture more rounded spiritualties in the nonisland, particularly urban com-

---

16. It is possible that further support of this kind may be found in the writings of Sorley MacLean, the Gaelic poet who was influential on the Macdonald brothers. See MacInnes 2006, 380–424.

munities. Lacking some of the distinctive interpretative influences that islanders experience, such theologies of place are often alien to these communities and spirituality is consequently a matter of the individual's or the community's relationship with God, lacking a sense of how that relationship connects with the place in which they dwell. A more knowing and explicit meditation on the aspects of the psalms discussed above and a more prominent use of the psalms in worship may counter such tendencies, though it may never have the force that it has for the islander.

Works Cited

Buchanan, Joni. 1996. *The Lewis Land Struggle: Na Gaisgich*. Stornoway: Acair.
"Listening Lounge." 2007. Ridge: Scotland's Independent Celtic Label. http://www.ridge-records.com/listen_salm.htm.
Macdonald, Calum, and Rory Macdonald. 2000. *Flower of the West: The Runrig Songbook*. Aberdeen: Ridge Books.
MacInnes, John. 2006. *Dùthchas nan Gàidheal: Selected Essays*. Edited by Michael Newton. Edinburgh: Birlinn.
MacLeod, John. 2008. *Banner in the West: A Spiritual History of Lewis and Harris*. Edinburgh: Birlinn.
Miller, Terry. 1984. "Oral Tradition Psalmody Surviving in England and Scotland." *Hymn* 35:15–22.
———. 2009. "A Myth in the Making: Willie Ruff, Black Gospel and an Imagined Gaelic Scottish Origin." *Ethnomusicology Forum* 18:243–59.
Parman, Susan. 2005. *Scottish Crofters: A Historical Ethnography of a Celtic Village*. 2nd ed. Belmont: Thomson Wadsworth.

Islands in the Sun:
Overtures to a Caribbean Creation Theology*

J. Richard Middleton

The Caribbean is a region of tremendous natural beauty. It is the sort of beauty that leads Harry Belafonte to poetically address his ideal (though unnamed) homeland, "Oh, island in the sun," and to promise that, "all my days I will sing in praise of your forest, waters, your shining sand" ("Island in the Sun," 1957).[1] Yet for all its undeniable natural beauty, the Caribbean is a region that is increasingly marred by pollution (for example, unsafe levels of toxins in fish in Kingston Harbor) and deforestation (for example, in Haiti, resulting in catastrophic mudslides in Gonaïves during heavy rains). So, the "forest, waters, [and] shining sand" of the pristine Caribbean are becoming more and more compromised by the human footprint. And this does not yet address the impact of natural disasters over which humans have no control, like Hurricanes Gilbert and Andrew in 1988 and 1992, respectively, or the devastating earthquake in Port-au-Prince in 2010.

The indelible human footprint on the natural beauty of the Caribbean—our impact on the earth—combined with horrendous natural disasters—the earth's impacts on us—gives the lie to any romantic vision of what we moderns have come to know as "nature," that is, the realm of the nonhuman. These ecological impacts also call into question the popular piety we find in the Caribbean church that imagines a separation between human "salvation," narrowly conceived, and our earthly environment. Paradoxically, among many Christians in the Caribbean and else-

---

* This is a revision of a chapter in Roper and Middleton 2013, 79–95; it is used by permission of Wipf & Stock Publishers.

1. Although Belafonte popularized the song, the lyrics were written by Irving Burgie (also known as Lord Burgess).

where, we find a decidedly otherworldly (and often individualistic) view of "salvation" as the saving of souls from a fiery judgment to spend an eternity with God in an ethereal heaven, combined with a romantic view of nature as a special place to encounter God. Witness the photographs on devotional greeting cards and posters and even the slides projected behind worship lyrics in some of our churches. Little if no thought is typically given to the possible connection—or, better, to the disconnect—between an otherworldly salvation and a romanticized nature.

There is no otherworldly salvation in "Islands in the Sun." What we find, rather, is an idyllic picture of nature joined to a naïve, almost primitive view of human society. Thus Belafonte sings, in the first verse, of "my island in the sun where my people have toiled since time begun." Apart from the hyperbolic lack of historical precision, since no people (not even the Amerindian Tainos and Caribs) have lived in the Caribbean from the beginning of time, the remainder of the song continues this romantic idealization of toil, whether it is the woman he sees "on bended knee cutting cane for her family," or the man he observes "at the water-side casting nets at the surging tide," or even the singer himself "lift[ing] [his] heavy load to the sky."

The juxtaposition of three verses, each mentioning physical toil or labor, with a fourth verse that fondly remembers drumming and Carnival, suggests that the "calypso songs philosophical" the singer mentions might function not to critique the social order, as much calypso has historically done, but as "philosophical" acceptance of the status quo. But perhaps it is not so much philosophical acceptance of the status quo as much as a positive *ignoring* of historical realities, as when the song's chorus states that this island in the sun was "willed to me by my father's hand." What world is or was the singer (or songwriter) living in, where the Caribbean is the natural inheritance of persons of African descent, without the intervention of European colonial powers? One searches this 1957 song in vain for any reference to the historical fact of colonialism or the history of European chattel slavery and later indentured labor, all of which decisively shaped Caribbean societies, these islands in the sun.

What are we to say of the present economic and social disparities in the Caribbean, fueled by the ideology and institutions of a global culture of consumerism? It is clear that the perspective of the song could not begin to address these contemporary issues. The issues are nonetheless important, and so this essay approaches those in the light of deep theological themes such as creation, redemption, imaging God, and salvation, within the contexts of Caribbean music, hymnody, and church practices.

Suspicion of Creation Theology

Caribbean theologians are right to express suspicions about any point of view that is blind to the reality of social inequities, especially if this blindness is combined with a romantic view of nature. When theologians and ethicists attempt to address the pressing needs of society, they often understandably focus on matters of human justice and injustice to the exclusion of significant reflection on the natural environment. There are certainly existential or pragmatic reasons underlying this suspicion of *creation* as a theme for theological reflection. Given the pressing human needs that face Caribbean people every day, it might seem that a theology of creation would take our focus off what is undeniably of prime importance.

But there is also a historical reason for the suspicion of creation as a theological topic. Theology as an academic discipline, both in the Caribbean and throughout the world, has been decisively shaped by a Western, Eurocentric habit of mind that distinguishes radically between *history* (people) and *nature* (the nonhuman). This distinction has its roots in the Renaissance split between freedom and nature, where thinkers like Pico della Mirandola (in his famous "Oration and the Dignity of Man") began to idealize human beings as transcending the determined and law-bound natural world (1978, 7–8), and it was fundamental to the rise of modern science in the sixteenth and seventeenth centuries in Europe, epitomized in Francis Bacon's quest for the seduction and conquest of nature by science (which illustrates well Susan Griffin's [2000, passim] contention that women and nature have been identified in Western thinking).

A theological version of the nature/freedom conceptual framework was given special momentum in the early twentieth century by Karl Barth, who famously distinguished immanent *religion* from transcendent *revelation*. To the former, explained Barth, God has pronounced a decisive *Nein!*[2]

While Barth himself was opposed to the hubris of modern Western humanism, the Barthian distinction between religion and revelation nevertheless contributed to a version of the history/nature distinction found

---

2. *Nein!* (No!) was the title of Barth's famous response to Emil Brunner's 1934 work entitled *Nature and Grace*, which itself interacted with Barth's earlier work. Brunner had proposed the validity of a creation theology (using the term "natural theology," though without the rationalism assumed by the classical tradition of that name). Both Brunner's proposal and Barth's response are published together in Brunner and Barth 1948.

in the biblical theology movement, associated with neoorthodox theologians like G. Ernest Wright. In the 1960s this movement tried to preserve the uniqueness of Old Testament revelation by contrasting debased Canaanite cyclical *nature* religion with the higher monotheistic linear, *historical* faith of the Bible (Wright 1943; 1950).[3] A version of this framework surfaces in the early works of Old Testament scholar Gerhard von Rad (1962, 137–39; 1984, 55–61), specifically in his claim that creation theology was a borrowing from Israel's pagan neighbors and in his refusal to allow that creation was integrally connected to Israel's salvation history or *Heilsgeschte*.[4] The history/nature dichotomy, without the overlaid value distinction, even shows up in Claus Westermann's (1978, 1–14; 1979, 28, 44) famous bifurcation between *salvation*—which is a matter of historical deliverance—and *blessing*—which is associated with matters like the birth of children and the fertility of flocks and land. This bifurcation simply cannot be sustained on exegetical grounds, since salvation in the Bible involves both deliverance from what impedes God's purposes and restoration to flourishing (Middleton and Gorman 2009, 45–46).

One particularly important version of the history/nature dichotomy is found in the prolific writings of Old Testament scholar Walter Brueggemann (1979, 161–85; 1988, 101–21; 2001, 39), who especially in his early works, programmatically claimed that creation faith (*nature*) served to justify the oppressive status quo both in Israel and its neighbors—the legitimation of order, he called it—while salvation/exodus faith (*history*) challenged the unjust ordering of the world in the name of a free and transcendent God.[5]

This complex theological inheritance may well constrain theologians to either prioritize a concern for human flourishing over a concern for the earth or to view creation theology with outright suspicion. But this anthropocentric focus, which separates human well-being from concern about the earth, is an artificial polarization, since people only exist, live, and work *somewhere*; that is, any sociocultural analysis would show that people both impact and are impacted by their environment. It is an arti-

---

3. See Middleton 2005, 186–88 for a discussion of the historical grounds as to why this distinction between Israel and the nations cannot be sustained.

4. Thankfully, von Rad (1972) later came to an appreciation of the role of creation theology in the Old Testament, evident in his mature work on the wisdom literature.

5. I have challenged Brueggemann's negative interpretation of creation theology in Middleton 1994.

ficial polarization from a biblical point of view as well, since humans are consistently understood in the Scriptures as part of the wider cosmos, which is not only created by God, but is the object of God's saving activity (Fretheim1993, 368–69; Middleton 2006, 86–91; Middleton and Gorman 2009, 45–54).

This is well understood by Caribbean theologian Ashley Smith, whose published works often address *creation* as the underlying basis of God's salvation in history. Especially in his seminal collection of essays, *Real Roots and Potted Plants: Reflections on the Caribbean Church*, we find a pervasive appeal to God as creator of the world and to God's purposes or intentions for creation as an alternative to a sacred/secular dualism and as a prod to appropriate ethical action on the part of Caribbean Christians (Smith 1984, 9–10, 15, 33–34, 38–39, 41, 44, 53, 65–66, 97–98, 100–101, 104).

Along these lines, Smith (1984, 15) claims that the church "needs to represent an attitude of affirmation in place of the traditional world-denial. To accomplish this, those who speak for it need to give greater prominence to the doctrine of creation." Or, as he puts it elsewhere, the key question before the Caribbean church is "what kind of ministry it might exercise at this particular time, in the name of him who continually makes all things new, in order that the purposes of his creation might be fulfilled" (44). He clearly states that to deny the goodness of the material world "is contrary to biblical teaching. It contradicts the Christian doctrine of creation and … goes against the New Testament understanding of the cosmic implications of the atonement (Romans 8:18–25).… The usual division of reality into sacred and secular is anything but Christian" (97).

I believe that Smith is on the right track in his theological appeal to creation to ground both salvation and ethics. We can no longer afford the luxury of suspicion about creation theology, that is, if we ever could have afforded that luxury. All places on earth are becoming more vulnerable to the global realities of climate change, toxic waste, overfishing, air pollution, and so on. Beyond the fact that the stresses humans are placing on the environment impinge on all peoples of the earth, including Caribbean people, it is also clear that our ecological crises are integrally connected to societal injustice. As James Cone (2001, 23), the father of black liberation theology in the United States puts it, "The logic that led to slavery and segregation in the Americas, colonization and apartheid in Africa, and the rule of white supremacy throughout the world is the same one that leads to the exploitation of animals and the ravaging of nature."

This is a profound observation, which should give us pause. Yet there is a something missing from Cone's analysis. I have no intention of devaluing Cone's historical and sociological approach, which is meant to challenge both ecological theologians and theologians who theorize race to take each other's work seriously. Yet one searches Cone's article, of which this is the opening statement, in vain, for any substantive theological or biblical analysis following from this important claim.[6]

Yet the Bible consistently interprets the connection between humans and the earth in a manner that positively contributes to a vision of human flourishing—at both individual and societal levels. The Bible is thus a powerful and often untapped resource on this topic. This suggests that the time is ripe for a *biblical* Caribbean theology that grounds human liberation in God's intent for creation and envisions a role for the earth within God's purposes.

However, this creation theology would need to move beyond professional theological interest in a public theology that addresses the large societal concerns of our times. Although such theological concerns are laudable—and even necessary—I believe that creation theology should be serviceable, not just for an elite cadre of Caribbean intellectuals, but for ordinary Caribbean Christians, to empower them in the universal priesthood of the believer that they might live with dignity, compassion, and power in a broken world, as a healing presence and witness to the coming kingdom of God through Jesus Christ.

These points are integrally connected, since the primary mode of access to theology for most Caribbean laypeople is precisely the Bible. We therefore need to develop a robust creation theology through a careful engagement with Scripture that would address the pressing need of ordinary Christians to internalize a vision of being human in God's world. Such a vision would integrally connect people and their societal needs to their bodies and their physical environment—and would connect salvation with God's creational intentions for this world.

In my reflections that follow, I address the otherworldly bent of much popular Caribbean Christianity by sketching the biblical teaching of the redemption of creation and by grounding this teaching in the

---

6. We find a similar problem in Leonardo Boff's (1997, esp. 104–114) important attempt at a rapprochement between liberation theology and ecology; although Boff does engage the question theologically, the Bible plays only a marginal role in the discussion.

Bible's affirmation of the earthly purpose of human life. It is the burden of this essay that a biblical creation theology addressed to Caribbean realities would both affirm the value and dignity of ordinary life and work in the world and would orient life and work toward God's larger redemptive purposes for justice and earthly flourishing.

The Bible's Vision of Cosmic Redemption

Central to the way the New Testament conceives the final destiny of the world is Jesus's prediction in Matt 19:28 of a "regeneration" (KJV, NIV) that is coming; Matthew here uses the Greek word παλιγγενεσία, which both TNIV and NRSV translate as "the renewal of all things," where the addition of the English phrase "all things" correctly gets at the sense of cosmic expectation.[7] Likewise, we have Peter's explicit proclamation of the "restoration [ἀποκατάστασις] of all things" in Acts 3:21, which contains the actual phrase "all things" (πάντων). When we turn to the Epistles, we find God's intent to *reconcile* "all things" to himself through Christ articulated in Col 1:20, while Eph 1:20 speaks of God's desire to *unify* or *bring together* "all things" in Christ. In these two Pauline texts, the phrase "all things" (τὰ πάντα) is immediately specified as things *in heaven* and things *on earth*. Since "heaven and earth" is precisely how Gen 1:1 describes the world God created "in the beginning," this New Testament language designates a vision of cosmic salvation, the redemption of the entire created order.

The cosmic vision expressed in other New Testament texts underlies the phrase "a new heaven and a new earth" found in both Rev 21:1 ("and I saw a new heaven and a new earth") and 2 Pet 3:13 ("we await a new heaven and a new earth in which righteousness dwells"; author's translation). The specific origin of the phrase "a new heaven and a new earth" is the prophetic oracle of Isa 65:17–25, which envisions a healed world with a redeemed community in rebuilt Jerusalem, where life is restored to flourishing and shalom after the devastation of the Babylonian exile (the phrase is found in Isa 65:17 and later in 66:22). The this-worldly prophetic expectation in Isaiah is then universalized to the entire cosmos and human society generally in late Second Temple Judaism and in the New Testament.

This holistic vision of God's intent to renew or redeem creation is perhaps the Bible's best-kept secret, typically unknown to most church

---

7. Unless otherwise specified, the biblical quotations that follow are from the NRSV.

members and even to many clergy, no matter what their theological stripe.[8] It is therefore particularly helpful to trace the roots of the New Testament vision in the Old Testament, in order to understand the inner logic of the idea.

The Human Calling to Image God on Earth

We should note that the Old Testament does not place any substantial hope in the afterlife; the dead do not have access to God in the grave or Sheol (Pss 6:5; 30:9; 88:3–5, 10–12; 115:17; Ecc 9:4–6, 10; Isa 38:9–12, 18).[9] Rather, God's purposes for blessing and shalom are expected for the faithful in this life, in the midst of history. This holistic perspective is grounded, theologically, in the biblical teaching about the goodness of creation, including earthly existence. God pronounced all creation including materiality good—indeed and at the end of the creative activity, "very good" (Gen 1:31)—and gave human beings the task to rule and develop this world as stewards made in God's image (Gen 1:26–28; 2:15; Ps 8:5–8).

In Gen 2:15 the original human task is to work and protect the garden, equivalent to agriculture. In Ps 8:5–8 humans are entrusted with rule over animal life on land, in air and water, the basis for the domestication of animals. And Gen 1:26–28 combines both agriculture and animal husbandry in its vision of humans created in God's image to rule animals and subdue the earth. Theodore Hiebert (1996, 42) is correct to note that, "in the preindustrial age of biblical Israel, it is impossible that the Priestly writer had more in mind in these concepts of dominion and subjection than the human domestication and use of animals and plants and the human struggle to make the soil serve its farmers."

In all these creation texts, the movement is "missional"—from God via humans outward to the earth. The paradox is of these texts is that the fundamental human task is both a matter of humble earthly service and yet a task of great dignity, namely, the responsible exercise of power on God's behalf in tending and developing the nonhuman world.[10]

---

8. This observation is based on my own experience in many different branches of the Christian tradition, both denominationally and theologically.

9. For analysis of these and other texts, see Wright 2003, 87–99.

10. This rule of the earth on God's behalf is precisely what Gen 1:26–28 means by the image and likeness of God (*imago Dei*), as is recognized by most Old Testament

It is sometimes shocking for readers of the Bible to realize that the initial purpose and raison d'être of humanity is never explicitly portrayed in the Bible as the worship of God, or anything that would conform to our notion of the "spiritual," with its dualistic categories. Instead, the Bible portrays the exalted human purpose in rather mundane terms of exercising power over our earthly environment as God's representatives. In the context of the ancient Near East, rule of the earth refers most basically to the development of agriculture and animal husbandry, which are the basis of human societal organization and ultimately leads to the development of all aspects of culture, technology, and civilization.[11] To put it another way, while Ps 148 and 96 call upon *all* creatures, including humans, to worship or serve God in the cosmic temple of creation (heaven and earth), the distinctive way *humans* worship or render service to the Creator is by the development of culture through interaction with our earthly environment in a manner that glorifies God. That is our fundamental human calling.[12]

By our communal development of culture through interaction with the earth and its creatures, humans function as God's image (*imago Dei*), mediating God's presence from heaven, where the Holy One is enthroned, into the earthly realm—as God's authorized and delegated representatives. By our faithful imaging of God through the ordinary, everyday tasks of human life (work, education, the raising of children, etc.), the human race was intended to bring the earth to its intended destiny as an integral part of God's cosmos-temple, filled with the divine presence and glory.

But in the biblical narrative a complication or impediment prevents completion of the original human purpose. Humans have misused the power God has given them, rebelling against their creator and turning

---

scholars. For an account of the history of interpretation of humanity as *imago Dei*, see Middleton 2005, ch. 1.

11. For further analysis of the human purpose in Genesis, see Middleton 2005, chs. 2 and 5; Crouch 2008, ch. 6; Cosden 2006, ch. 4; Wolters 1999, 27–33; and Middleton and Walsh 1995, ch. 6.

12. There are two important points to make here. First, the cultural development of the earth, rather than "worship" narrowly conceived, is explicitly stated to be the human purpose in biblical texts recounting the creation of humanity. "Worship" in the narrow sense may be understood as *part of* human cultural activity. Secondly, we should not reduce human worship/service of God to verbal, emotionally charged expressions of praise, which is what we usually mean by the term "worship." Note that Paul in Rom 12:1–2 borrows language of sacrifice and liturgy from Israel's cult in order to describe full-orbed bodily obedience which, he says, is our true worship.

against each other. This misuse of the power of *imago Dei* is manifest most fundamentally in disobedience toward the Creator (Gen 3), which then blossoms into a pattern of violence and fractured relationships among people (Gen 4–11), which continues to this day. Whereas the early chapters of Genesis do, indeed, record the continuing cultural development of the earth—including the first city (4:17) and the development of nomadic livestock herding, technology, and music (4:20–22)—we also have the first murder (4:8) and Lamech's revenge killing (4:23–24), till violence fills the earth (6:11). The biblical tradition understands that human transgression of God's norms leads to death, which is the antithesis of God's purposes for earthly flourishing; our contemporary predicament is that death in its manifold forms has invaded and degraded human life and the entire earthly creation.

Salvation as the Restoration of God's Purposes for Creation

The biblical affirmation of earthly life is articulated in the central and paradigmatic act of God's salvation in the Old Testament, the exodus from Egypt. Israel's memory of this event testifies to a God who intervenes in the harsh realities of history in response to injustice and suffering. But more than that, the exodus is presented as not just *from* bondage, but *to* or *for* shalom under a new and beneficent overlord. This new time of shalom is attained when the redeemed are settled in a bountiful land and are restored to wholeness and flourishing as a community of justice living according to God's wise laws.

In line with the creational grounding of salvation, Old Testament legal and wisdom literature reveals an interest in mundane matters such as the fertility of land and crops, the birth of children and stable family life, justice in the city, and peace in international relations. The Old Testament does not spiritualize *salvation* but understands it as God's deliverance of people and land from all that destroys life and the consequent restoration of people and land to flourishing (Fretheim 1993, 371–72; Middleton and Gorman 2009, 45, 47, 52–54). And while God's salvific purposes narrows for a while to one elect nation in their own land, this "initially exclusive move" is, as Old Testament scholar Terence Fretheim (2005, 29) puts it, in the service of "a maximally inclusive end," the redemption of all nations and ultimately, the entire created order.

Although the Old Testament initially does not envision any sort of positive afterlife, things begin to shift in some late texts. Thus in Ezekiel's famous vision of the valley of dry bones (Ezek 37), the restoration of Israel is portrayed using the metaphor of resurrection, after the "death" they suffered in Babylonian exile. But this is arguably still a metaphor, not an expectation of what we would call resurrection. Then, a protoapocalyptic text like Isa 25:6–8 envisions the literal conquest of death itself at the messianic banquet on Mount Zion where God will serve the redeemed the best meat and the most aged wines. This Isaiah text anticipates the day when YHWH will "swallow up death forever," cited in 1 Cor 15:26, 54, and "wipe away all tears," echoed in Rev 21:4. But the most explicit Old Testament text on the topic of what we would call resurrection is the apocalyptic vision of Dan 12:2–3, which promises that faithful martyrs will awaken from the dust of the earth to which we all return at death, according to Gen 3, to attain "eternal life."

It is important to note that this developing vision of the afterlife has nothing to do with "heaven hereafter." The expectation is manifestly this-worldly, meant to guarantee for the faithful the earthly promises of shalom that death has cut short. The Wisdom of Solomon 3 is particularly helpful here. This text specifically associates "immortality" with reigning *on earth* (Wis 3:1–9, esp. 7–8), that is, resurrection is a reversal of the earthly situation of oppression, the domination of the righteous martyrs by the wicked, which led to their death, and thus is the fulfillment of the original human dignity and status in Gen 1:26–28 and Ps 8:4–8, where humans are granted rule of the earth.[13]

These ancient Jewish expectations provide a coherent theological background for Jesus's proclamation of the kingdom of God. Jesus construes the kingdom of God as "good news" for the poor and release for captives (Luke 4:18), which he embodied in healings, exorcisms, and the forgiveness of sins, thus reversing particular distortions of earthly life. These expectations also make sense of Jesus's teaching in the Sermon on the Mount that the meek would "inherit the earth" (Matt 5:5) and later in Matthew that "at the renewal of all things," the cosmic "regeneration," the

---

13. Contrary to the misreading of Leonardo Boff (1997, 79–80), the use of the metaphor of "rule" either in the Bible or in contemporary theology does not automatically legitimate unlimited dominion or exploitation of the earth. Rather, "rule" is an ancient way to speak of the exercise of power, which may be beneficent or destructive. It is used in Second Temple Jewish tradition to dignify human life, often in situations of oppression. Jesus himself suggests (and models) that the normative exercise of rule is humble service of others (Mark 10:42–45).

disciples would reign and judge with him on thrones (Matt 19:27–30). This helps us understand Rev 5:9–10, which envisions a redeemed church from "every tribe and language and people and nation" constituted as "a kingdom and priests to serve our God, and they will reign on the earth." Also Rev 22:3–5 indicates that when God's throne, which is currently in heaven is finally established on earth, God's servants will "reign forever." The word "forever" disabuses us of the idea this might be some sort of temporary millennium to be followed by an otherworldly eternal state. Rather, what Revelation offers is the eschatological restoration of the original human calling as *imago Dei* to administer and develop this world to God's glory.

The eschatological restoration taught by Jesus and envisaged in Revelation has begun in the church, which is even now being renewed in the image of God (Eph 4:24; Col 3:9–10) to become the "new humanity" (a much better translation than "new self," which we find in most modern translations).[14] This means that day-to-day sanctification is a matter of the restoration of our humanness, with all that entails, as we are called to live up to the stature of Christ, whose perfect imaging becomes the model for the life of the redeemed (Phil 2:5–11; Eph 4:13). The day will come when we are fully conformed to the likeness of Christ (1 John 3:2), which will include the resurrection of the body (1 Cor 15:49).

So when Paul describes Jesus's own resurrection from the dead as the "firstfruits" of those who have fallen asleep (1 Cor 15:20), he claims that the harvest of new creation has already begun, the expected reversal of sin and death is inaugurated. This reversal will be consummated when Christ returns in glory climactically to defeat evil and all that opposes God's intent for life and shalom on earth (1 Cor 15:24–28). Then, in the words of Rev 11:5, "the kingdom of this world [will] become the kingdom of our Lord and of his Messiah." At that time, explains Paul, creation itself, which has been groaning in its bondage to decay, will be liberated from this bondage into the same glory God's children will experience (Rom 8:19–22)—that is, the glory of resurrection.

The inner logic of this vision of holistic salvation is that the Creator has not given up on creation, but is working to salvage and restore the world, human and nonhuman, to the fullness of shalom and flourishing intended

---

14. The KJV has "the new man." This term portrays regeneration as corporate, not just individual.

from the beginning. And redeemed human beings, renewed in God's image, are to work towards and embody this vision in their daily lives.

The Otherworldly Hymnody of the Church

The tragedy is that this kind of holistic vision of salvation is found only rarely in popular Christian piety or even in the liturgy of the church. Indeed, it is blatantly contradicted by many traditional hymns and contemporary praise songs sung in the context of communal worship. This is an important point since it is from what they sing that those in the pew or auditorium typically learn their theology, especially their eschatology.

From the classic Charles Wesley hymn, "Love Divine, All Loves Excelling" (1747), which anticipates being "changed from glory into glory / till in heaven we take our place," to John Thomas McFarland's "Away in a Manger" (1904–1908), which prays, "And fit us for Heaven, to live with Thee there," congregations are exposed to and assimilate an otherworldly eschatology.[15] Some hymns, like James M. Black's "When the Roll Is Called up Yonder" (1893), inconsistently combine the idea of resurrection with the hope of heaven:

> On that bright and cloudless morning when the dead in Christ shall rise,
> And the glory of His resurrection share;
> When His chosen ones shall gather to their home beyond the skies,
> And the roll is called up yonder, I'll be there.

Some hymns even interpret resurrection without reference to the body at all, such as Thomas Shepherd's "Must Jesus Bear the Cross Alone?" (1693), which in one stanza regards death as liberation: "Till death shall set me free." Another verse of the hymn asserts: "O resurrection day! / When Christ the Lord from Heav'n comes down / And bears my soul away" (added by Henry Ward Beecher, 1855).

A hymn like "When We All Get to Heaven" may be too obvious, but notice that George Bernard's "The Old Rugged Cross" (1913) ends with the words, "Then He'll call me some day to my home far away / Where his

---

15. Many hymns have different authors for different stanzas; in each case I have indicated the author for the particular stanza quoted and the known or estimated date of composition or first publication.

glory forever I'll share," and "Just a Closer Walk with Thee" (anonymous, early twentieth cent.) climaxes with the lines:

> When my feeble life is o'er,
> Time for me will be no more;
> Guide me gently, safely o'er
> To Thy kingdom shore, to Thy shore.

Likewise, Christian H. Bateman's "Come Christians, Join to Sing" (1843) affirms that "On heaven's blissful shore, / His goodness we'll adore, / Singing forevermore, / 'Alleluia! Amen!'"

This notion of a perpetual worship service in an otherworldly afterlife is a central motif in many hymns. For example, William R. Feathersone's "My Jesus I Love Thee" (1864) affirms that "In mansions of glory and endless delight, / I'll ever adore Thee in heaven so bright." Similarly, William C. Dix's "As with Gladness Men of Old" (ca. 1858) asks in one stanza that "when earthly things are past, / Bring our ransomed souls at last / Where they need no star to guide" and in another stanza expresses the desire that "In the heavenly country bright / ... There forever may we sing / Alleluias to our King!"

Thankfully, most hymnals no longer have the sixth verse of John Newton's "Amazing Grace" (1773), which predicts,

> The earth shall soon dissolve like snow,
> The sun forbear to shine;
> But God, who called me here below,
> Will be forever mine.

Yet Chris Tomlin's contemporary revision of this classic hymn, known as "Amazing Grace (My Chains Are Gone)" (2006), reintroduces this very verse as the song's new climax, ready to shape the otherworldly mindset of a fresh generation of young worshipers unacquainted with hymnals.

This overview of hymns just scratches the surface of worship lyrics that portray the final destiny of the righteous as transferal from an earthly, historical existence to a transcendent, immaterial realm. As the theologian and preacher, A. W. Tozer, is reputed to have said, "Christians don't tell lies; they just go to church and sing them."[16] Perhaps that is too

---

16. This quote is found all over the Internet, without an explicit citation to Tozer's works. Noted Tozer scholar James L. Snyder admits that while it may not be found in

harsh; nevertheless, I can testify to the steady diet of such songs that I was exposed to, growing up in the church in Kingston, Jamaica, which certainly reinforced the idea of heaven as otherworldly final destiny.

Echoes of Creation Theology in Caribbean Music

I am, however, perpetually grateful that along with such exposure I came to know, through sheer proximity, the this-worldly theology of Rastafarianism, especially as mediated through the music of Bob Marley and the Wailers. While I am a committed Christian and thus cannot affirm everything found in Rasta theology, I nevertheless discern a deeply rooted biblical consciousness in the lyrics of many Wailers's songs.[17] For example, the song "We an' Dem" claims that "in the beginning Jah created everythin' / and he gave man dominion over all things," and "Pass It On" asserts that "In the kingdom of Jah / man shall reign."[18] These lyrics express, in admittedly androcentric language, the biblical vision of this-worldly dignity granted humans at creation, a dignity that will be restored in the kingdom of God.

And Peter Tosh's version of "Get Up, Stand Up," a song he cowrote with Marley, understands well the implications of a creation-oriented eschatology for ethics, when it contrasts the doctrine of the rapture with a desire for justice on earth:

> You know, most people think,
> A great God will come from the skies,
> And take away every little thing
> And lef' everybody dry.
> But if you know what life is worth,
> You would look for yours
> Right here on earth

---

a specific published work, the quote accurately echoes what Tozer has said in some of his sermons (available in audio recordings); "it is Tozer and it expresses his feelings on the subject" (personal communication, December 20, 2010).

17. I have explored the theology of a number of songs by Bob Marley, Peter Tosh, and Bunny Wailer in Middleton 2000, 181–204.

18. *Jah* is the shortened form of the divine name YHWH (Yahweh/Jehovah) found in expressions like "hallelujah!" (which literally means "praise YHWH!"). Rastafarians love to quote Ps 68:4 in the KJV: "Sing unto God, sing praises to his name: extol him that rideth upon the heavens by his name JAH, and rejoice before him."

And now we see the light,
We gonna stand up for our rights.[19]

The song goes on to critique the "preacher man" for taking the focus off earthly life and affirms that the singer is "Sick and tired of this game of theology, / die and go to heaven in Jesus name." This is the very theology that leads Marley, in the song "Talkin' Blues," to admit, "I feel like bombing a church, / now that you know that the preacher is lying." But if Tozer is right, it is not just the preacher who is lying, but also the worshipers who blithely sing hymns of escape to an ethereal heaven, when the Bible teaches no such thing.

Practical Implications for the Caribbean Church

Yet, the preacher without a doubt bears the larger share of guilt. As Jas 3:1 warns, "Not many of you should become teachers, my brothers and sisters, for you know that we who teach will be judged with greater strictness." Here the culpability of Caribbean Christian leadership is evident. For it is the mandated responsibility of church leaders to teach "the whole purpose of God" (Acts 20:27), rather than some truncated version of this purpose. Of course, the otherworldly orientation of popular Caribbean theology could easily and legitimately be blamed on our colonial past, since we learned this theology from our European colonizers. But to shift the burden of responsibility to others would be to let ourselves off too lightly. The Caribbean church must engage in serious self-examination and come to terms with the fact that its own leaders have perpetuated an escapist theology that entrenches ordinary Christians still further in despair and paralysis, as they pine for a heavenly home distant from the everyday realities of Caribbean life.

Historically, the otherworldly vision that has been inculcated into the consciousness of the Caribbean church allows for little or no explicitly Christian norms to guide life in contemporary society, with the prominent exception of sexual mores. In particular, an otherworldly focus on heaven hereafter prevents the biblical gospel from addressing the economic and

---

19. These lyrics are transcribed from Tosh's *Equal Rights* album (1977); the song first appeared on the Wailers's *Burnin'* album (1973) with slightly different lyrics. The lyrics are different again on Tosh's *Captured Live* album (1984) and on Bunny Wailer's *Protest* album (1977).

societal realities of our time. Thus when Caribbean preachers begin to speak, as they are now doing, to the genuine need to overcome poverty among their congregations, their preaching often echoes the idolatrous greed and selfishness of Western consumer culture, baptized with a thin veneer of Christian language.

While it is laudable to motivate church members to move beyond acquiescing in poverty, the so-called prosperity gospel that is gaining ground in the Caribbean church is a betrayal of the biblical vision of *shalom*, which ought to direct the church towards communal care for neighbors and the earth. The point is that simply casting out the old demons of otherworldliness, without an engagement with a truly biblical spirit or ethos, allows the wandering spirits of the age—unclean spirits—to inhabit our souls. Today the Caribbean church is in danger of buying into the worst elements of consumerist individualism at the heart of Western culture.

We, therefore, need a radical reformation in the Caribbean, in both the teaching and worship of the church. I propose that if the church's teaching and worship were grounded in a biblical creation theology that addresses earthly concerns in a holistic manner, this would have the potential to guide the church's life in the contemporary world in at least three ways.

First of all, a biblical creation theology can provide a foundation and orientation for the value and holiness of daily life as we live out our identity as *imago Dei* in society. This identity, as it is renewed in Christ, obliterates the artificial split between "sacred" and "secular" and gives meaning to the mundane challenges of life, work, family, and education, interpreted as the outworking of our sacred calling to be human in God's world. As Smith so eloquently put it,

> The enlightened or awakened Christian who is aware of the biblical doctrines of creation and redemption is liberated from the fallacy that there is a part of the world which is outside of the sphere of God's activity and his love, and therefore, inherently, unholy and under condemnation. Being aware of the holiness of all creation and of God's concern in all that happens in the world, the Christian ... participates wholeheartedly, joyfully and responsibly in all the affairs of his community. (Smith 1984, 38–39)

Secondly, biblical creation theology can provide an ethical challenge to the present unjust and corrupt status quo. Understanding the biblical vision of God's original intent for life on earth can allow us to discern a world out of whack with how things were meant to be. Creation theology thus provides the church with a critical principle of dissent from the injustice in the world, so that we do not simply baptize the present as God's will.

Finally, a biblical creation theology provides an empowering vision of God's purposes for shalom that can energize church members both as individuals and in community to utilize their gifts and opportunities to make a difference in the world by how they live. A church that has its eyes firmly fixed on the coming of God's kingdom from heaven to earth, rather than on leaving earth for heaven, will seize the moment and seek to contribute to healing, justice, and earthly flourishing in the whole range of human life and activities. In this way, the church in the Caribbean may grow into a living foretaste of the coming of God's kingdom to this our beautiful yet broken and needy earthly home.

Works Cited

Boff, Leonardo. 1997. *Cry of the Earth, Cry of the Poor*. Translated by Phillip Berryman. Ecology and Justice. Maryknoll, NY: Orbis Books.

Brueggemann, Walter. 1979. "Trajectories in Old Testament Literature and the Sociology of Ancient Israel." *JBL* 98:161–85.

———. 1988. *Israel's Praise: Doxology against Idolatry and Ideology*. Philadelphia: Fortress.

———. 2001. *The Prophetic Imagination*. Rev. ed. Philadelphia: Fortress.

Brunner, Emil, and Karl Barth. 1948. *Natural Theology*. Translated by Peter Fraenkel. London: Geoffrey Bles.

Cone, James H. 2001. "Whose Earth Is It, Anyway?" Pages 23–32 in *Earth Habitat: Eco-Justice and the Church's Response*. Edited by Dieter Hessel and Larry Rasmussen. Minneapolis: Fortress.

Cosden, Darrell. 2006. *The Heavenly Good of Earthly Work*. Milton Keynes, UK: Paternoster; Peabody, MA: Hendrickson.

Crouch, Andy. 2008. *Culture Making: Recovering Our Creative Calling*. Downer's Grove, IL: InterVarsity Press.

Fretheim, Terence E. 1993. "Salvation in the Bible vs. Salvation in the Church." *WW* 13:363–72.

———. 2005. *God and World in the Old Testament: A Relational Theology of Creation*. Nashville: Abingdon.

Griffin, Susan. 1978. *Woman and Nature: The Roaring Inside Her*. New York: Harper & Row.

Hiebert, Theodore. 1996. "Re-Imaging Nature: Shifts in Biblical Interpretation." *Int* 50:36–46.

Middleton, J. Richard. 1994. "Is Creation Theology Inherently Conservative? A Dialogue with Walter Brueggemann." *HTR* 87:257–77.

———. 2000. "Identity and Subversion in Babylon: Strategies for 'Resisting Against the System' in the Music of Bob Marley and the Wailers." Pages 181–204 in *Religion, Culture and Tradition in the Caribbean*. Edited by Hemchand Gossai and N. Samuel Murrell. New York: St. Martin's Press.

———. 2005. *The Liberating Image: The* Imago Dei *in Genesis 1*. Grand Rapids: Brazos.

———. 2006. "A New Heaven and a New Earth: The Case for a Holistic Reading of the Biblical Story of Redemption." *JCTR* 11:73–97.

Middleton, J. Richard, and Brian J. Walsh. 1995. *Truth Is Stranger than It Used to Be: Biblical Faith in a Postmodern Age*. Downer's Grove, IL: InterVarsity Press.

Middleton, J. Richard, and Michael J. Gorman. 2009. "Salvation." *NIDB* 5:45–61.

Pico della Mirandola, Giovanni. 1965. *Oration on the Dignity of Man*. Translated by A. Robert Caponigri. Chicago: Regnery.

Rad, Gerhard von. 1962. *The Theology of Israel's Historical Traditions*. Vol. 1 of *Old Testament Theology*. Translated by D. M. Stalker. New York: Harper & Row.

———. 1972. *Wisdom in Israel*. Translated by James D. Martin. Nashville: Abingdon.

———. 1984. "The Theological Problem of the Old Testament Doctrine of Creation." Pages 55–61 in *Creation in the Old Testament*. Edited by Bernhard W. Anderson. Philadelphia: Fortress.

Roper, Garnett, and J. Richard Middleton, ed. 2013. *A Kairos Moment for Caribbean Theology: Ecumenical Voices in Dialogue*. Pittsburgh: Pickwick.

Smith, Ashley. 1984. *Real Roots and Potted Plants: Reflections on the Caribbean Church*. Mandeville, JM: Eureka.

Westermann, Claus. 1978. *Blessing in the Bible and in the Life of the Church*. Translated by Keith R. Crim. OBT. Philadelphia: Fortress.

———. 1979. *What Does the Old Testament Say About God?* Edited by Friedemann W. Golka. Atlanta: Knox.

Wolters, A. M. 1999. "The Foundational Command: 'Subdue the Earth!'" Pages 27–32 in *Year of Jubilee, Cultural Mandate, Worldview*. Edited by B. van der Walt. Study Pamphlet 382. Potchefstroom, ZA: Institute for Reformational Studies.

Wright, G. Ernst. 1943. "How Did Early Israel Differ from Her Neighbors?" *BA* 6:1–21.

———. 1950. *The Old Testament against Its Environment*. SBT 2. Chicago: Regnery.

Wright, N. T. 2003. *The Resurrection of the Son of God*. Christian Origins and the Question of God 3. Minneapolis: Fortress.

The Island of Tyre: The Exploitation of Peasants in the Regions of Tyre and Galilee

Hisako Kinukawa

Identifying myself as an islander, I cannot evade thinking of the imperialism, colonialism, and nationalism of my country, Japan, especially in relation to other Asian countries. Geographically, the land of Japan, consisting of four main islands surrounded by almost four thousand small islands, is separated from other nations by the ocean. The nearest country, Korea, is at least ninety-three miles across the Japan Sea, which is usually very rough with storms and seasonal typhoons. Because of these geographical advantages, Japan has been little exposed to the threat of being conquered by other countries. In addition, the country is favored by a mild climate with four seasons.

Since 80 percent of the total land of Japan is covered by mountains, agriculture has been limited to small parts of its plains and hills. It is well-known that part of the reason for Shogun Hideyoshi to send his army in 1592–1598 to the Korean peninsula was to colonize it as well as to get food supplies. There was also a geopolitical problem. The strongest reason for Hideyoshi's invasion operations was his ambition to negotiate with the then empire of Spain that had controlled and colonized most of Asia, locating its strategic front in Luzon in the Philippines. He needed to show the strength of his reign against Spain and prevent any further invasion.

Ethnically, there has been a myth, or an illusion, that the Japanese people are a homogeneous race. This myth has given rise to the belief that it is important to maintain the purity of Japanese blood. The concept of ethnic homogeneity has been identified with superiority to other people—especially in Asia—connected with the religious concept of purity and used by the authorities to exploit people belonging to other races and ethnicities.

Before and during the Second World War (1909–1945), Japan invaded Asian countries, colonized them, and devoured even their people as if they were commodities. After the war, as the national economy rose, it devoured agricultural as well as sea products from Asian countries. It employed and paid people an extremely low wage.

The spirit of insularity in Japan has most explicitly been reflected in its expansionism, gobbling up products and people in Asian countries, which has given me a distinct perspective as I interpret a story with which I struggled for a long time: the story of the Syro-Phoenician woman. In the Gospel of Mark, the story revolves around a tiny island in the Mediterranean Sea. This island was the affluent trading center of Jesus's time. In this chapter, my focus is to ask what impact the island had on the story.

The Region of Tyre, the City of Tyre

Mark sets the story of the Syro-Phoenician woman (7:24–30) on an occasion when Jesus "went away to the region of Tyre" (v.24). Using the term "region" for Tyre, Mark apparently separates the region of Tyre from the city of Tyre.[1] This introduction raises some questions. Where is the city of Tyre located? How is it distinguished from the region of Tyre? Why did Mark carefully mention that Jesus went to the *region* of Tyre instead of the city of Tyre? Is this Mark's deliberate intention? If so, why?

The woman in the story is introduced as "a gentile, of Syro-Phoenician origin." She heard of Jesus, when "Jesus set out and went away to the region of Tyre" (7:24), where she must have resided.[2]

It has been contested whether Jesus actually traveled to the region of Tyre. He might not have been there, but the phrase "region of Tyre" seems to play a critical role for interpreting the whole story. First, the region of Tyre intersected the region of Galilee, with no clear borders separating the two.[3] We may plausibly imagine that villages of different ethnic groups

---

1. Gerd Theissen (1991, 66. n.18) points out that "the preferred reading of the Textus Receptus and, most recently, H. Greeven, *Synopse der drei ersten Evangelien*, *methori* (= border, territory surrounded by frontiers), makes this still clearer. Another interesting reading is the peculiar *horeh* (Miniscule 565), reminiscent of the hill country on the border between Tyre and Galilee."

2. Unless otherwise stated, all biblical quotations are my own translation.

3. See Jonathan Reed (2000, 185–86), who uses archeological evidence to make this point.

were intermingled in the regions of Tyre and Galilee. Second, Syro-Phoenicia, a colonial state of the Roman Empire, reflects the political situation of Jesus's day and therefore suggests that Mark might have spoken from the same social location as the woman. She is introduced as ethnically Syro-Phoenician and politically and socioculturally as a Greek/gentile. Third, it seems that Mark avoided saying the *city* of Tyre when he refers to this woman. He may intend to point out that there is an extraordinary discrepancy between the city of Tyre and its region.[4]

Encounter of Syro-Phoenician Woman with Jesus

She came to him because she had a demon-possessed little daughter. She came to meet Jesus with a plea for the healing of her sick daughter. Daring to cross the borders of her culture and possibly her religion[5] shows that she was on the verge of a desperate situation with her daughter.

Despite her fervent plea, however, Jesus rebuffed her with unexpectedly harsh words. "Should you allow the children to be fed first, for it is not fair to take the children's food and throw it to the little dogs?" (7:27). He explicitly and totally rejected her plea through words that despised her people as dogs and defended his own people as children.

Nevertheless, she was not knocked down by Jesus's rejection. She talked back to him. "Yes, it is so, but, sir, even the dogs under the table eat the children's crumbs" (7:28). She had not wanted to lose her self-respect by seeking help from a foreigner, but put herself at the mercy of the foreigner for the sake of her sick daughter. She therefore kept her subordinate stance that she might maintain a relationship with him. We learn that gaining a concrete answer for her plea from him was for her a death-and-life issue. This foreigner was her last hope.

---

4. Theissen gives literary evidence on the region of Tyre by referring to Josephus's writings. He also writes, "This restriction to the countryside is all the more astonishing because, at the time when the synoptic tradition was being shaped and the Gospels written, there was already a Christian community in Tyre (cf. Acts 21:3-6), and it would have been natural either to connect Jesus with the city or to associate the residents of the city with Jesus. The restriction of Jesus to the rural area may thus correspond to the real pre-Easter situation" (1991, 66–67).

5. From the coin they produced, we could tell that the main god of the people in Syro-Phoenicia was Melkart.

My Questions

Since I met this woman in Mark, I have kept asking two particular questions. First, why did Jesus use such harshly negative words against the woman's plea? Second, why did he use the image of table fellowship to deny her plea for the healing of her sick daughter? Talking about food instead of dealing with sickness sounds out of place in this context.

Though I have read more than seventeen interpretations offered by feminist exegetes and theologians on the story, I did not find one article giving clear clues to the two questions I raised.

Where Is the City of Tyre?

Then I took a detour to ask where the city of Tyre is. Almost all the maps offered in the Bibles I consulted or even in map books on biblical cities and towns hardly show the geography of Tyre. Tyre is shown as a tiny dot that looks like a part of the land of current Israel/Palestine on the east shore of the Mediterranean Sea.

Reading the book of Luke-Acts, we encounter the following text, showing the route of Paul's third trip along the coast of the Mediterranean Sea. Having set sail from Ephesus,

> we came by a straight course to Cos [island], and the next day to Rhodes [island], and from there to Patara. When we found a ship bound for Phoenicia, we went on board and set sail. We came in sight of Cyprus [island]; and leaving it on our left, we sailed to Syria and landed at Tyre [island], because the ship was to unload its cargo there. (Acts 21:1–3)

There are seven geographical names referred to in this small text. The Greek text does not say which ones are islands, maybe because the audience of the text was well-acquainted with the geography. As far as I have been able to discover, no English translation mentions which names designate an island, but this short message mentions four islands, including Tyre. We, those who are distant from the ancient Mediterranean world, have so little knowledge about them that it is especially hard for us to recognize that the city of Tyre used to be an island, because when we check "Tyre" on the current Google Earth, it looks like the tip of a small peninsula.

The City of Tyre on an Island

Gerd Theissen's *The Gospels in Context* offered me my first perspective of the city of Tyre as an island (1991, 72–75). Historical and geographical study shows that the city of Tyre had surely been on an island five hundred meters apart from the mainland. It flourished through trading, and its fortress had been known as impregnable. It was Alexander the Great (356–323 BCE) who, having built a path between the seashore and the island, attacked and finally captured the island of Tyre (Barnett et al. 2014).

Therefore, although by the time of Jesus the city of Tyre was no longer on an island, the city still must have been affluent as the trading center of the Mediterranean world. It is not clear when the island was actually connected to the mainland after being conquered by Alexandar the Great.

The City of Tyre in the Hebrew Scripture

The place name Tyre is mentioned often in the Hebrew Scriptures. Tyre is said to be proud and a threat to Israel. It is always named with Sidon (Isa 23; Jer 47:4; Ezek 27, 28; Joel 3:4–8; Zech 9:2) and described as polluted by materialism.[6] In chapters 27 and 28 of his book, Ezekiel describes in great detail how Tyre and Sidon were corrupted by materialism.

Theissen points out that the city of Tyre, the center of trading in the Mediterranean world, was affluent because of its "world-wide" import/export business.[7] Ezekiel's report in 27:3 ("When your wares came from the seas, you satisfied many peoples; with your abundant wealth and merchandise you enriched the kings of the earth," NRSV) shows how well the city of Tyre was flourishing as a world trading center.[8] It "catalogs the vast

---

6. In the story of the Syro-Phoenician woman, some important manuscripts, including Sinaiticus, Alexandrinus, and Vaticanus, add "and Sidon" after "Tyre." See O'Day 1989, 291.

7. See Theissen (1991, 72–73). He writes, "Tyre was a rich city, its wealth based on metal work, the production of purple dye (cf. Pliny *Nat. Hist.* 5.17.76; Strabo *Geogr.* 16.2.23), and an extensive trade with the whole Mediterranean region. Its money was one of the most stable currencies in circulation at this period; it survived for decades without significant devaluation. This was certainly one reason why the temple treasury was kept in Tyrian coin, even though this meant accepting the fact that coins of Tyre depicted the god Melkart."

8. These trading centers (in Ezek 27 and 28) include, from west to east: Tarshish (Sardina, or Tartessus in Southern Spain), which traded silver, iron, tin, and lead;

extent of its commerce, covering most of the then-known world" (Barnett et al. 2014). Products of Egypt, Spain, Babylonia, Arabia, and the lands of the Euphrates and Tigris Rivers were coming and going through Tyre to other parts of the world. Tyre played an invaluable role as a transit center of produce.

The Region of Tyre

In contrast, the region of Tyre was on the mainland with poor soils. Most of the people in the villages of Tyre were peasants and suffered from poor products. Even worse, most products were taken away to feed the affluent people in the city of Tyre. Theissen writes about how dependent the island of Tyre was on the agricultural products of the regions of Tyre as well as Galilee.

> Tyre had a problem: its rural territory was limited by natural factors. Tyre was on an island, and the nearby strip of coast suitable for farming was narrow. For agricultural products, the city depended on imports, and this *dependence was a constant in the city's history*. We find *indications* of this situation even in the Old Testament: Solomon sent wheat and oil to Hiram of Tyre (1 Kgs 5:25) [which would have been around 950 BCE].... According to Ezekiel, Judea and Israel sent wheat, olives and figs, honey, oil, and balm to Tyre (Ezek 27:17). No wonder that a

---

Javan (Ionians, i.e., Greeks); Tubal (people settled in current Southern Turkey, east of Anti-Taurus mountains); and Meshech (Assyrian "Mushku," west of the Anti-Taurus mountains), which traded human beings and vessels of bronze; Beth-Togarmah (Assyrian "Tilgarimmu," in current central Turkey, east of the southernmost Halys River, east of Tobal), which traded horses, war horses, and mules. From south to north, Helbon (thirteen miles north of Damascus), which traded wine and white wool. And from southwest to northeast: Uzal (modern Sana in Yemen), which traded iron, cassia, sweet cane; Dedan (west central Arabia), which traded saddlecloths; Sheba (southwest Arabia), which traded spices, precious stones, gold; Heran (on the Balikh River in Mesopotamia); Eden (Assyrian Bit-Adini), Canneh (southeast of Haran); Asshur (south of Nineveh); Chilmad (unidentified Mesopotamian city), which traded choice garments, clothes of blue and embroidered work, carpets of colored material; Rhodes (a big island in the Aegean Sea), which traded ivory tusks and ebony; Edom/Aram, which traded turquoise, purple embroidered work, fine linen, coral, and rubies; Judah and Israel, which traded wheat, millet, honey, oil, and balm; and Kedar, which traded lambs, rams, goats.

drought in Palestine also led to famine in the region of Tyre and Sidon (1 Kgs 17:7–16). (Theissen 1991, 73, emphasis added)[9]

Historical Glance on Tyre

Let us take a look at the history of Tyre so that we may understand the economic discrepancy between the city of Tyre and its regions. The Phoenicians arrived in the land of Phoenicia around 3000 BCE. They had kings claiming divine origin, supported by a council of elders.[10] However, the kings' power "was limited by that of the merchant families, who wielded great influence in public affairs" (Barnett et al. 2014). This gives evidence of the affluence and influence of those who were in the trade industry. During and under the protection of the Eighteenth Dynasty of Egypt (ca. 1550–1300 BCE), the Phoenicians developed their trading business. The invention of an alphabet by a Semitic people in Egypt facilitated their trading business and improved communication among traders. Within a short time the use of an alphabet spread to the parts of the world where trading ships voyaged. The Phoenicians contributed to the advance of the use of an alphabet in a much wider world, as far west as the current Spain, through their trading business (see Boeree 2014).[11]

Barnett et al. (2014) detected a wide range of their activities, noting a "fresco in an Egyptian tomb of the 18th dynasty depicting seven Phoenician merchant ships that had just put in at an Egyptian port to sell their goods, including the distinctive Canaanite wine jars in which wine, a drink foreign to the Egyptians, was imported."[12]

---

9. Theissen adds, "Josephus writes still more plainly that Solomon every year sent 'grain and wine and oil' to King Hiram 'of which, because … he inhabited an island, he was always particularly in need' (*Ant.* 8.141; cf. 8.54)."

10. Barnett et al. (2014) also report that "Tyre's first colony, Utica in North Africa, was founded perhaps as early as the tenth century BCE," which shows the geopolitical strength of Tyre supported by merchants.

11. George Boeree reports a discovery of rock carvings in southern Egypt that indicates that people began using the alphabet at least in the 1900s BCE, contrary to the earlier theory of its beginning in the 1700s BCE.

12. With regard to Phoenician exports and imports, Barnett et al. write: "The exports of Phoenicia included cedar and pine woods from Lebanon, fine linen from Tyre, Byblos, and Berytos, cloth dyed with the famous Tyrian purple (made from the snail Murex), embroideries from Sidon, metalwork and glass, glazed faience, wine, salt, and dried fish. The Phoenicians received in return raw materials such as papy-

The Phoenicians are also known for their prominent skill in navigation on the oceans. They are also "credited with the discovery and use of Polaris (the North Star). Fearless and patient navigators, they ventured into regions where no one else dared to go, and always, with an eye to their monopoly, they carefully guarded the secrets of their trade routes and discoveries and their knowledge of winds and currents" (Barnett et al. 2014; see also "Lebanon: Tyre"). However, their prosperity and power made them enemies, by whom the city experienced a variety of sieges in its history.

In 332 BCE Alexander the Great (Macedonia) set out to conquer this strategically valuable island for a base of international trading. Unable to beat the city, he cut off the supply of food for seven months. Tyre held on. But the conqueror forced the people of the abandoned mainland to build a causeway, and once within reach of the city walls, Alexander concentrated his power of army forces to batter and finally fell the fortifications. In 64 BCE, "Phoenicia was incorporated into the Roman province of Syria, though Aradus, Sidon, and Tyre retained self-government. Tyre continued to mint its own silver coins" (Barnett et al. 2014).

Thus the city of Tyre, being tossed around by various powers rising one after another,[13] survived more or less as the center of trading in the ancient Mediterranean world. We may plausibly conclude that at the time of Jesus the city of Tyre, which may have been no longer an island but was now connected with the land, was still flourishing in its trading business. The people living in the city of Tyre may be considered affluent elites, though they had to depend on the agricultural products brought in from the villages of Tyre (Phoenicia) and Galilee (Judea). Peasants in the villages had a hard time raising their crops in the poor soil and under the exploitative pressure granted by the Roman Empire, the Herodian monarchy, and the temple-centered authorities of religion.

---

rus, ivory, ebony, silk, amber, ostrich eggs, spices, incense, horses, gold, silver, copper, iron, tin, jewels, and precious stones" (Barnett et al. 2014). Barnett et al. add that "the Phoenicians also conducted an important transit trade, especially in the manufactured goods of Egypt and Babylonia. From the lands of the Euphrates and Tigris, regular trade routes led to the Mediterranean. In Egypt the Phoenician merchants soon gained a foothold; they alone were able to maintain a profitable trade in the anarchic times of the 22nd and 23rd dynasties (c. 950–c. 730 BCE)" (Barnett et al. 2014).

13. There were at least five sieges before the time of Jesus: 724–720 BCE, Assyrian siege by King Shalmaneser V; 705 BCE, Assyrian siege by King Sennacherib; 663 BCE, Assyrian siege by King Ashurbanipal; 586–573 BCE, Babylonian siege by King Nebuchadonezzar II; and 332 BCE, Macedonian siege by Alexander the Great.

Discrepancy between the City and the Region of Tyre

Most of the produce was purchased by rich city dwellers, while the peasants in the villages were always in want. Galilean peasants must have been resentful when they saw their ruling class selling their produce to the highest bidders from urban Tyre. Agricultural crops produced by the peasants did not return to their daily table to satisfy their own basic needs. The peasants experienced a constant shortage of food and money, even though they labored from dawn to dusk all through the year. Under the exploitation of the three parties oppressing them, the Galilean peasants were deprived of a stable life.

Taking into consideration the bitter economic relationship between the affluent city of Tyre and exploited Galilee, we can see that Jesus's bitter words thrown to the woman would have had a powerful impact in one sense. The saying, even though it is so offensive to the woman, reflects the humiliating power relationship that Galileans had to endure with respect to urban Tyrians. The words could mean, "First, let the mouths of the poor people in Galilee be satisfied. For it is not good to take poor people's food and throw it to the rich Tyrians in the city." The words overtly express the reality of the destitute Galilean peasants and show their strong resistance against the power exercised by the urban people of Tyre. Those Tyrians who hungered for and devoured the agrarian produce of Galilee become the "dogs" in Jesus's analogy. Jesus's reply may represent the popular feeling of the Galilean peasants toward the Tyrians, whom they viewed as rich and representing the Hellenistic culture of the elite.

Understandably, we can see why Jesus's response to the woman's plea was expressed in terms of table fellowship. Maybe we could say Jesus's only concern was about Galilean peasants having little food for their table. When he had no thought other than to rebuff her, his mind was only with his people in Galilee. Though he was physically facing her, he seemed not interested in knowing who she was and what she actually sought. She was still the "other" before him.

What Impact She Gave Jesus

From the historical, political, and social backgrounds of villages of Galilee and Tyre we have already seen, we may say the woman with the demon-possessed daughter is from one of the peripheral villages of Tyre, where people's lives are not as easy as the lives of those in the urban city. The two must have been socially ostracized because of the demon possession and therefore had to live in a peripheral village. She must be neither rich nor privileged. That might be the reason she could persevere the negative words thrown by Jesus.

We may say she had the courage and nerves of a protective mother to hang on with Jesus. If she was alone by herself, that is, if her need was only for herself, she might not have come to him. Her tenacity enabled her to talk back to Jesus saying, "Yes, it is so, but, sir, even the dogs under the table eat the children's crumbs" (7:28). First, she stood herself on Jesus's side and acknowledged the primacy that the Galilean peasants ought to have. She admitted unfair distribution under the dominant relationship of the urban Tyrians over the Galileans. We should notice it was not he but she that crossed the borders first, and it was also she who invited him to cross them, as an expression of island hospitality.

It seems she is raising a serious question here for Jesus. That is, why can he totally ignore a sick child while talking about feeding his "children," the others of Galilee? She is insisting that Tyrians are not monolithic, and she is one of the others in the society of Tyre. She asks Jesus to be consistent in putting the primacy on the marginalized wherever they are. It seems like she is asking Jesus to come to her side and see how desperate her situation is. Had she not experienced being the "other" in her society, she would not be able to be as confident as she is in asking Jesus's help. Her tenacity asks his egalitarian spirit to work regardless of race, sex, and state. She dares to ask Jesus to expand his circle of table fellowship to any destitute person.

It was her tenacity in defending her daughter and herself that made him turn around, see, and finally accept her as a person. Her words must have had a powerful impact such that he seemed to be shocked and then totally changed his words into those announcing the healing of her daughter. "For your words, you may go" (7:29). He fully accepted her request.

Reflection

The two questions I raised in the beginning are answered by inquiring into the history of the city of Tyre and its solid affluence with the poorly developed agriculture in the regions of Tyre and Galilee.

It is reasonable for us to imagine what might have happened in Jesus's mind. In the beginning, he did not show any interest in her as a person. His mind was fully occupied with the horrible situation under which his people had to suffer. However, her tenacious challenge shocked him and resulted in his crossing a border of ethnicity. This should deserve a special mention.

It may not have actually happened for him to go beyond the political or social border between the two societies, but her challenge showed him his movement should have no border as long as it sticks to the most powerless people. Her desperate request for justice of her life has made him come out of his own social location and move his mind into another social location. We can never make light of the woman and her challenge even to shake the heart of Jesus.

It is not a welcome or comfortable awareness that I am from a country of islands, because the experience of islanders, surrounded by the water, inevitably asks us to expand our horizon to the social and historical relationships with other parts of the world, beyond ourselves, especially neighboring countries. As stated in the beginning, the history of my country cannot be engaged without mentioning the atrocities of devouring other people's territories and products. Being aware of this part of the history of Japan, I received new eyes to think about my long-time question on Jesus's harsh statements against the Syro-Phoenician woman. As I already mentioned, his dismissive statements were, representing the suffering peasants in the regions of Galilee and Tyre, at the mercy of the devouring elites in the city part of Tyre

Works Cited

Barnett, Richard D., Glenn Richard Bugh, Samir G. Khalaf, Paul Kingston, Clovis F. Maksud, and William L. Ochsenwald. 2014. "Lebanon." *Encyclopaedia Britannica Online*. http://www.britannica.com/EBchecked/topic/334152/Lebanon.

Boeree, C. George. 2014. "The Origin of the Alphabet." *The Evolution of Alphabets Online*. http://webspace.ship.edu/cgboer/alphabet.html.

"Lebanon: Tyre." MiddleEast.com. http://www.middleeast.com/tyre.htm.

O'Day, Gail R. 1989. "Surprised by Faith: Jesus and the Canaanite Women." *List* 24.

Reed, Jonathan L. 2000. *Archaeology and the Galilean Jesus: A Re-examination of the Evidence*. Harrisburg, PA: Trinity Press International.

Theissen, Gerd. 1991. *The Gospels in Context: Social and Political History in the Synoptic Tradition*. Translated by Linda M. Maloney. Minneapolis: Fortress.

Sea-ing Ruth with Joseph's Mistress*

Jione Havea

A story [*talanoa*] is like a river. And like a river it trickles from the source until it flows flows flows. Down the mountains of the mountains. Branching onto the land the land the land. Flowing. Spiralling. Flowing towards the sea. Spiralling towards the sky. Where it grows wings and flies towards the universe of the unknown. (Figiel 1999, 3–4)

Sea-ing Talanoa

Stories gather at the sea of *talanoa* (story, telling, conversation), then wing to the unknown, in flight, leaving behind their roots, like the widowed Ruth leaving her mother's home in Moab, and the usual, like Joseph in flight, leaving his garment in the hands of Potiphar's wife. In Oceania, *talanoa* have rhythms (Vaka'uta 2011) that ripple (Havea 2010) looking for companion imaginations and memories, like the pilot fish that attach to the underside of sharks and whales. They ride and feed, find home and security, and i suspect that they in return tickle their hosts in special ways.[1]

In the sea of *talanoa*, a story is a companion for, and commentary on, other stories. One tells, remembers, hears, and reconfigures a story in relation to, like or unlike, poking and pulling, other stories. Stories draw near to, open toward, other stories, luring, seducing, and trimming one another when they meet. In this regard, stories do not finish and cannot be

---

* A presentation made at the Research Seminars of the School of Theology, Charles Sturt University (16 Mar 2012) gave birth to this essay and benefited from the careful reading and comments by Margaret Aymer.

1. I prefer the lowercase "i" because i use the lowercase with "you," "she," "they," and "others"; i do not see the point in capitalizing the first person when she or he is because of everyone/everything else.

finished off. They shift (Havea 1996) between here and there (Havea 1995), in terms of both time and space.

Appreciating a story as one ripple in a sea of *talanoa* is one island view (see Havea 2008) that problematizes the romanticizing view that isolates islands and islanders, as in the demeaning paralleling of "no story is an island" with "no text is an island." Such sayings emphasize how islands are small and not connected to larger lands to which islands and islanders are considered useless.[2] This is why, Spivak (2008, 9–10, 248) explains, the Pacific (Oceania) continues to be "absent" in Asia-Pacific. The Pacific is that part of the Asia-Pacific region that scholars prefer to fly over. True, islands are small even if islanders think and talk big. Size matters in some intercourses, but size does not determine value in *talanoa* cultures. *Talanoa* is about impact (see Havea 1998). Whether long or short, a *talanoa* (story, telling, conversation) ripples when it is sharp and poignant. Like sex, *talanoa* is about climax, sometimes real, sometimes faked.

The ones who say that no story or text is an island do not know what it is like to be islanders. Islands are distant and hidden from larger lands, but being out of sight does not mean disconnection. Indeed, islanders are relational people. Our ancestors spoke of their home islands as "floating lands"—which the locals in Onotoa, Kiribati, call *Aba tabebeiti*—and they crossed between islands quite frequently even before European explorers dared to enter our seas (Havea 2015). Islanders have a different sense of space. The watery sea is real space that is not off limits. Distance is not barring, and home is place of transit. Islander navigators perceive islands moving toward them, instead of their rafts moving toward land. In these regards, island living has to do with transiting,

---

2. This is evident in the comment by analyst Zhixing Zhang, who is reported to say concerning Hilary Clinton's visit to Australia and New Zealand in 2010: "This is the bit that I don't understand, why does anyone want to counter Australian and New Zealand dominance of Polynesia/anything east of the Australian/NZ coast? It doesn't have population, its resources are tiny … and its position is not very strategic in nature. Australia and NZ are the jewel of Australasia, the islands are hardly anything at all and all you'd take Australia for is resources and to deny other nations from using it as an FOB [forward operating base]/surveillance point with which to push up from the south. And even then all you have to do is hold Indonesia/Melanesia and you've blocked that route anyway. I just don't get why anyone gives a shit about Polynesia" (Hubbard and Hager 2012).

drifting, floating, changing, constantly, and accordingly *every story, text, and talanoa is an island*.

Every story, text, and *talanoa* is a floating land, a drifting home, a routing memory, and a rooting desire in a sea of stories. The fluid world of *talanoa* is where stories cross each other's paths, inter-course and in transit. In the sea of stories, a story transtexts (Havea 2003, 1–10), undresses (Joseph), and uncovers (Boaz), and appreciating it is no longer just about its birthing context and intended meanings but also about its crossing with other stories, as guided by [pilot fish] readers. I, an openly piloting and fishy reader, imagine many stories poking other stories with a demand similar to that laid upon Joseph: "Lie with me" (Gen 39:7b).

The Hebrew Bible is a sea of stories also. It contains more than stories, for it also has instructions, poems, songs, sayings, proverbs, parables, and more, but i read those as currents in the biblical sea of stories. Whereas narrative critics prefer to read a story in its entirety (see Fewell and Gunn 1993a), following the plot and the development of characters (so Fewell and Gunn 1993b), readers of the Hebrew Bible have a habit of jumping back and forth between stories, pursuing a host of links and allusions (see Fewell 1992). Selectivity flows in the veins of readers, and such a tendency is encouraged by the fragmenting rotation of lectionary readings that churches follow (see Havea 2009). These latter days, not too many people read the whole Hebrew Bible or a whole book of the Hebrew Bible.

In Oceania, no one swims in the whole deep sea (*moana*) or tells every story in our sea of *talanoa*. Islanders too are selective, intentional about the *talanoa* we tell and interweave. This essay is my first attempt to *sea* the stories of two women, Ruth the Moabite widow and the nameless mistress of the house that Joseph served, popularly identified as wife of Potiphar, but i refer to her as "mistress of Joseph." As wife of the master (boss), she is Joseph's mistress. There is obvious sexual overtone, and so in the following reading i will pursue the hint that she was lover also to Joseph. My interest in these two characters, i confess, is because they are cultural outsiders to the dominating Israelite flow of the Hebrew Bible and because their *talanoa* are bodily and spicy. The crossing of these two points of interest raises a question that is poking but which i will not seek to unravel in this essay: Are the *talanoa* of these women sexualized because they are cultural outsiders (especially in terms of heritage, feature, color and language) as in the cases of Hagar, Rahab, and Jezebel also?

Finding a Partner for Ruth

The book of Ruth has received varied readings (see Greenstein, 1999) and several partners in the history of its interpretation (see Eskenazi and Frymer-Kensky 2011, lvi–lxx). The most interesting pairing is with the book of Esther. The only books in the Hebrew Bible named after women characters, one Jewish (Esther) the other Moabite (Ruth),[3] prospering in a foreign land. There are differences between the main women characters, for instance, Ruth was a migrant and a widow while Esther was diaspora-born and received attention as a virgin. Nonetheless, partnering these two books and women characters is interesting to me, because it is done by people who favor different interpretive lenses and interests.

Though McGee (1988) and Bush (1996) read within the Baptist traditions, they come with different agendas to Ruth and Esther. McGee's interest is on the subject of redemption, and he looks into the story of Ruth for the kinsman-redeemer (*go'el*) feature. Because of his interest with redemption, McGee turns to the story of Esther, in which he sees God working behind the scenes to bring redemption to the Jewish people. God is not named in Esther, but McGee finds God in the narrative, through Ruth and the concern for redemption. And although McGee sees a connection between the two stories, he treats them separately, as two different books, in a two-section monograph.

Bush offers a commentary in the traditional understanding of that genre, seeking to locate the biblical texts in their literary, historical, and sociopolitical milieu, then comments and explains what those texts meant to say and mean (back then and there). Bush does not explain why he has brought Ruth and Esther under the same cover; his commentary too is a two-part work.

Linafelt and Beal (1999), who have contributed to developing narrative and postmodern criticisms in biblical studies, offer a different kind of commentary. They take a book each, Linafelt engages Ruth while Beal tackles Esther, but they pretty much, so to speak, keep the two women in different rooms. The page numbering is indicative of the separation upheld—the page numbering of the commentary on Esther restarts instead of continu-

---

3. "The book of Ruth is the only biblical book bearing the name of a gentile. This fundamentally important point is in itself already a sign of subversion" (LaCocque 2004, 1).

ing the page numbering of the commentary on Ruth. The two stories are brought to the same sea, but their paths are not seen to cross.

The collection of essays that Brenner edited (1999) is intentional about crossing the two. Finally! Brenner's work is in three sections (I. Ruth; II. Ruth and Esther: Mothers and Daughters; III. Esther), and the contributors are openly feminist. The middle section consists of only two chapters, but those set the stage for actually partnering Ruth with another biblical character.

Fewell and Gunn (1999), following literary allusions, also point Ruth (and Naomi) toward other biblical characters: Lot and his daughters, Judah and Tamar, Rachel and Leah, and others. Other commentators bring other biblical characters alongside Ruth, and the affirmation of reading Ruth intertextually with other texts (see Eskenazi and Frymer-Kensky 2011/571, xxi–xxvi) crosses with the ripples of sea-ing *talanoa* in the previous section.

I join this conversation with another partner for Ruth, this time an Egyptian woman who is not named but who wields influence in her story. Their partnership is the upshot of this *talanoa* (telling). I will attend to them in turns, engaging with one in the shadows of the other.

Uncovering Ruth

Ruth is the Moabite widow who accompanied her widowed mother-in-law, Naomi, as she turns back to her homeland, Bethlehem, in Judah. Naomi was returning home with neither husband nor sons, while Ruth was leaving her mother's house and her homeland. Ruth had a co-sister-in-law, Orpah, who opted to turn back to Moab.[4]

Ruth's story is kindled with movements, breaking away and returning home, circling around the lack and plenty of food, beginning and ending in Bethlehem, "house of bread," which suffered famine then became prosperous with grains after YHWH took note of the people and visited, returned to, the land again (Ruth 1:6). YHWH departs, then visits; seeds desert, then returns; people go, then come. Bodies migrate from land to land and from life to death, seeking survival, accompanied by the migration of cultures and languages and the politics of immigration (Honig 1999). The

---

4. Naomi started the journey with Orpah and Ruth (Ruth 1:6–7), then at a point on the way she told them to "return" to Moab (1:8). I imagine that they were outside of Moab, and on their way, when Naomi told them to turn back.

narrator of Ruth's story does not deal with the struggles due to cultural and language differences or with the burdens of migration, nor does the narrator of Joseph's story, but as a migrant myself i take note of those matters especially when i read stories of migration and border crossing.

Migration is a strong flow in the biblical sea of *talanoa*. The chief ancestor of the Israelites, Abram, was in transit in Haran, on the way from Ur of Chaldea, in Babylonia, when YHWH told him to go to Canaan (Gen 11:31–12:4a). He and his family were already on the way to Canaan when YHWH called him to go; he was called while he was in transit. Migration takes place for a variety of reasons in the biblical sea of *talanoa*, with famine as a key cause (see, e.g., Gen 12:10; 42:1–5). Negotiation of cultural and language differences is more difficult for refugees, with women and children struggling the most. Gender matters in the experience of migrants. Ruth would therefore have been vulnerable in Judah, more so than Joseph would have been in Egypt.

In Bethlehem, across the Dead Sea from Moab, Ruth is a foreigner with strange customs and a different language world of thoughts and practices. She would have stood out and most likely received unfair treatment, for it is common to mistreat strangers who are different from the dominant group: different color, different features, different accent, different treatment. There is, on the other hand, a tendency to romanticize the foreigner, and i imagine that Ruth would have received such attention also from some of the locals. She would have been loathed and cherished, prized and bothersome, all because she was a foreigner, someone who did not fully belong. A childless widow upon arrival to Bethlehem, some would have seen her as cursed, or barren, as Bethlehem was (famine) at the beginning of the story and Egypt would be while Joseph was there.

When Ruth reached Boaz's field (Ruth 2:2–3), she came as an enigma (especially to those who did not know her). The young man in charge of the field introduced her to Boaz not as a widow but as a "Moabite young woman" whom Naomi brought back with her (2:6), and Boaz greets her as "daughter" (2:8) and assures her that she would be safe and cared for well in his field (2:9). He has lots of grains, and it would be to her advantage to glean only at his field. Like most migrants, Ruth was conscious that she was a foreigner and felt that she did not deserve being singled out (2:10). It turned out that Boaz heard about her previously. The two of them were not strangers to each other even if this was the first time they meet. At the end of the day, Ruth was loaded with grains when she returns to Naomi. They agreed that she would continue to glean at Boaz's field up to the end of the harvests, with a detour to the threshing floor. From the field she moves inside, like Joseph moving into his mistress's chamber.

When Ruth came to the threshing floor, she was driven. Naomi instructed her what to do:

> So bathe, anoint yourself, dress up, and go down to the threshing floor. But do not disclose yourself to the man until he has finished eating and drinking. When he lies down, note the place where he lies down, and go over and uncover his feet and lie down. He will tell you what you are to do. (Ruth 3:3-4; NJPS)

The direction is clear, and Ruth submits. Naomi's plan was for Ruth to corner Boaz into providing a home where Ruth would be happy (Ruth 3:1). Home for Ruth, so far, means home for Naomi also. The two women are one. Ruth carries out Naomi's plan with stealth. Two birds will eat the victim of one stone.

Boaz was contented when he goes to lie down at the end of the heap of grain. He may have gone to lay there in order to guard his seeds. She approaches, uncovers his feet, and lies down. In the middle of the night, Boaz startles and pulls back. "Behold, a woman was lying at his feet" (Ruth 3:8).

The story does not leave much room for the imagination (contra LaCocque 2004, 81-82), and it is difficult to beat around the bushes. A woman who uncovers the feet of a sleeping man drunk with merriness and lies down is not looking for a soft spot to sleep out the evening. At the very least, she is looking to make the sleeping man think that he had sex with her (so Fewell and Gunn 1999, 237-38). Since she is a widow who appears to have completed her mourning period (cf. Gen 38), i imagine that she might have wanted to have sex. In other words, she wanted more than appearing to have sex. She wanted to have sex, and Boaz responded in his sleep. He startled. Imagine. He stirred. What in him stirred? What of him startled?

Ruth was at his feet, and Boaz was grateful: "You are blessed of YHWH, my daughter. Your latest loving kindness is greater than the first, *for you did not go after younger men*, whether poor or rich. Now, my daughter, do not be afraid. All that you say, I will do to you for all the elders know *what a vigorous woman you are*" (Ruth 3:10-11, emphasis added). Ruth pins Boaz to submission, and he does all the talking. Boaz takes over the role of Naomi, giving Ruth instructions:

"Lie down until morning" (3:13).

"Let it not be known that the woman came to the threshing floor" (3:14).

"Hold out the shawl you are wearing" (3:15).

Ruth leaves the threshing floor with more seeds from Boaz. Back to the domain of Naomi, the wiser older widow knows the impact of what happened the previous night: "Stay, my daughter, till you learn how the matter turns out. For the man will not rest, but will settle the matter today" (Ruth 3:18). Settlement is urgent, as it is also in the case of Potiphar's wife.

As another migrant, i admire Ruth's shrewdness and celebrate her courage. Her gender and foreignness did not stop her from grabbing security for her and her mother-in-law and from forcing Boaz to act on her behalf. The story continues on to the gathering of elders at the town gate where Boaz made the rightful redeemer hand over his sandal and the rights to the inheritance of Elimelech, including Ruth, but such would not have taken place if Ruth did not uncover Boaz's feet. Nothing would have happened had she been shy and naïve, as many migrants are. She was driven, inspired, and purposeful.

There is something praiseworthy about Ruth's action, in the name of solidarity (with Naomi), for the sake of survival and belonging, demanding (Boaz) action and accountability, in summary, seizing the night, the space, and the opportunity that came with stirring Boaz. She performed with no need of the guilt-trip with which Mordecai ambushed Esther, "For if you keep silence at such a time as this, relief and deliverance will rise for the Jews from another quarter, but you and your father's family will perish. Who knows? Perhaps you have come to royal dignity for just such a time as this" (Esther 4:14).

Ruth of Moab, who was widowed when her Judean husband died prematurely, acted in her own interest, fulfilling her desires and assuring her survival. I affirm and applaud her for those. She entered the story as a partner and desirous woman (wife); i affirm her as a desiring and sexual being in the biblical sea of *talanoa*.

Undoing Mistress

Nameless, the woman character in Gen 39 is identified in relation to her husband, Potiphar, who is master of the house. She is not a subject in her own rights but a possession of her husband. She is Mrs. Potiphar, who became "a caricature of the foreign temptress who tries to seduce the righteous Joseph" (Jeansonne 1990, 107–8). A woman, an Egyptian, her gender and ethnicity is a double whammy in the patriarchal eyes of pro-Israelite readers. She is the outsider in the pro-Israelite mindset, but Joseph is the foreigner in the narrative.

Namelessness suggests a lack of respect. The narrator does not care enough about a character to remember her or his actual name. On the other hand, being nameless is somewhat freeing. The nameless character cannot be pinned down to a particular label; she or he cannot be named or blamed. She or he might be the possession of another character to whom blame and praise could be directed, but as far as the nameless character is concerned, she or he is free of specificity or particularity. This line of reasoning will insult in modern contexts where the Cartesian self is favored and tutored, but welcomed in circles where relations are more important than individuality. Namelessness is not a problem in a world where, as the southern African philosophy of Ubuntu upholds, "I am because of who we are." A subject is always in relation to, because of, others. I can therefore be nameless and not be troubled.[5]

The namelessness of the female character in Gen 39 opens her up to alternative markings of identity. She is the lady of the house, spouse of the master, mistress of the household. There is no indication if Potiphar had another wife, so she comes across as one with higher status than the other people, male and female, free and slave, in her household. She is a woman with power. She is the Naomi in this *talanoa*. She speaks with authority: "Lie with me" (39:7). "By using this abbreviated speech, the narrator suggests that the woman's passion is intense and lustful" (Jeansonne 1990, 110). Unlike Ruth, this mistress is not into stealth. She demands and expects satisfaction. Joseph refuses, giving up the chance to redeem her.

The text is ambiguous whether Potiphar gave her also into the care of Joseph: "So [Potiphar] left all that he had [in house and field, see Gen 39:5b] in Joseph's charge; and with him there, *he had no concern for anything but the food that he ate*" (39:6a, emphasis added). In this, the narrator suggests that she too was committed into Joseph's hand. The narrator could have been literal, so "food" really means "food," but Joseph invites

---

5. From the side of my comrades, my namelessness is a problem, and they often seek to draw me out of my hiddenness.

a metaphorical reading when he equates himself to Potiphar: "He is not greater in this house than I am, *nor has he kept back anything from me except yourself, because you are his wife*" (39:9a, emphasis added). In this connection, "food" means "wife," as "feet" also means "genitals" in Ruth's story. In this turn, Joseph is both proud (he equals his master) and pious (he will not touch his master's food/wife). Ruth gleans grains; Joseph cleans house only.

Whether the woman was given to Joseph or not, her demand of him is straight and unambiguous. She wanted Joseph, who is described as "handsome and good looking" (Gen 39:6b; NRSV)—or as Alter suggestively translates it, "comely in features and comely to look at" (Alter 2004, 222)—to lie with her. He is so handsome that in the Qur'an addition (Q Yusaf 12:30–34) partying women cut their hands upon seeing him saying, "Allah preserve us! This is no man—this is a noble angel!"

> Biblical narrative rarely comments on the physical features of characters, but when it does so those physical qualities usually play an important role in the story. In this case the mention that Joseph is handsome helps to explain why his master's wife is attracted to him. (Kaltner 2003, 25)

There is no beating around the bushes here, no waiting for Joseph to fall asleep. The mistress is direct. She demands with economy of effort, using only two Hebrew words.

> Against her two words, the scandalized (and perhaps nervous) Joseph will issue a breathless response that runs to thirty-five words in Hebrew. It is a remarkable deployment of the technique of contrastive dialogue repeatedly used by the biblical writers to define the differences between characters in verbal confrontation. (Alter 2004, 222)

Many readers find fault with her aggressiveness. She is read as a sex-hungry woman with an indecent reputation, and she is hard to like (Kaltner 2003, 42). I, on the other hand, admire her, as i did Ruth, because she knew and went for what she wanted. She is a strong woman. So i turn back to ponder other ways of making sense of this story.

Randall C. Bailey has suggested in *talanoa* (conversation) that the text hints that a sexual relationship between Potiphar and Joseph was possible, based on the identification of Potiphar as officer or courtier (*saris*) of Pharaoh (39:1). In Hebrew and Aramaic, *saris* is the word usually translated as eunuch (a detail that Scullion [1986, 59] considers a late addition and Sarna [1989, 271] ignores). For Bailey, a eunuch who gives everything he has to his handsome attendant is suggestive. Potiphar could have given himself also into Joseph's hands. This is the sublime suggestion of Joseph's admission that Potiphar gave him everything, except his wife. Joseph got Potiphar, but not his wife.

Drawing from Bailey's reading, i turn to Joseph's mistress. Because she was married to a eunuch, she lacked the opportunity to be seeded. The eunuch might be able to "get it up," but he could not impregnate her. It thus makes sense to me that she wanted Joseph to lie with her. She wanted Joseph to be for her in the way that her husband could not (so Jeansonne 1990, 109). The text invites this reading, and this is manifested in Alter's translation, which presents the text as an open can of worms:

> And she laid out his garment by her until his master returned to his house. And she spoke to him things of this sort, saying, "The Hebrew slave came into me, *whom you brought us, to play with me*. And so, when I raised my voice and called out, he left his garment by me and fled outside." And it happened, when his master heard his wife's words which she spoke to him, saying, "Things of this sort your slave has done to me," he became incensed. And Joseph's master took him and placed him in the prison-house, the place where the king's prisoners were held. (Gen 39:17–20, emphasis added)

Her charge ripples in the currents of ambiguity. On the one hand, she charges that Joseph came into her space (room, chamber), as Ruth came to Boaz's corner of the threshing floor, but she could be understood also, on the other hand, that she is referring to sexual encounter (penetration). Joseph did come into her, and coming, entering, בא, is one of the Hebrew expressions for sex. In this regard, meaning slides, and the sliding of meaning is not foreign to this story. In Gen 39:7, her "lie with me" is clear. But in 39:10, the narrator fudges: "And so she spoke to Joseph day after day, and he would not listen to her, *to lie by her*, to be with her" (Alter 2004, 223, emphasis added). The narrator softens her proposition, from the demand to "lie with me" to "lie by her." Expressions shift, and meanings slide.

Because there is no punctuation in Hebrew, there is room for the possibility (so Alter 2004, 224) that Potiphar brought Joseph into his house so that he would play with her: "The Hebrew slave came into me, [the one] whom you brought us to play with me." I imagine something similar to what happened to Sarai, whom Pharaoh took *as his wife* (Gen 12:10–

20; see also Gen 20:1–18, 26:1–16). In this connection, Potiphar brought Joseph "to play" with his wife (also).[6] The problem then is not that she forced herself onto Joseph, but that he refused to play with her. In refusing, as did the Redeemer at the gate of the elders (Ruth 4:6), he insults his mistress (boss, lover).

The Qur'an brings the Lord to Joseph's rescue and accelerates the flow of the narrative (compared to the biblical account):

> She locked the doors and said, "Come here." He responded, "Allah forbid!" (12:23b) ... She desired him and he would have desired her if not for the clear proof of his Lord. (12:24a)

> They ran to the door and she ripped his shirt from behind. They met her husband at the door. She said, "There is no penalty for a man who desires to do evil to your family other than imprisonment or painful punishment." He responded, "She tried to entice me!" A witness from her family testified, "If his shirt is torn in the front, she is telling the truth and he is a liar. But if his shirt is torn from behind, then she is lying and he is truthful." When he saw that his shirt was torn from behind he said, "This is one of your plots. Truly, your plots are great. Ignore this, Joseph. You, woman, ask forgiveness for your offense. Truly, you are a sinner." (12:25–29)

The Qur'an version puts a door in the way of Joseph's flight.

> Symbolically, the door represents a border that Joseph is not allowed to cross. When the woman locks it and makes his passage impossible, Joseph becomes a prisoner in her world, unable to escape. This sense of entrapment is heightened when he rushes to the door only to meet her husband, who is yet another potential barrier to his freedom. (Kaltner 2003, 35)

Rippling Talanoa

In the sea of *talanoa*, stories drift, ripple, and, the Samoan novelist Sia Fiegel suggests, fly to the unknown. This evokes for me a Tongan expression, *puna*

---

6. Compare with Qur'an 12:21, "The Egyptian who bought him said to his wife, 'Treat him well during his lodging. Perhaps he will be of benefit to us and we will take him as a son.'"

ki he ta'e'iloa,⁷ flight to the unknown, which is about mobility, migration, transit, courage, faith, the course of which this essay crosses. This essay however, so to speak, has been a piloted flight. I piloted the *talanoa* of Ruth and of Joseph's mistress toward each other. The upshot is that Ruth and Joseph's mistress are no longer strangers, unknown, *ta'e'iloa*, to one another. Something like this happens in the *moana* (deep sea) of *talanoa*.

This essay manifested two features of *talanoa*: telling (*talanoa*) two stories (*talanoa*) together. In the sea of *talanoa*, the stories of Ruth and of Joseph's mistress are no longer isolated from one another. The island practice of *talanoa* unmoored the two stories and their characters from their literary and historical harbors so that they drift toward and lie with one another. The third feature of *talanoa* (conversation) remains at this point (due to the limits of the medium of writing) as an invitation, for readers to engage and respond to the reading offered herein. *Talanoa* does not come to rest with the stories nor with their telling, but it invites readers and listeners to respond, engage, and interact. In this regard toward kindling *talanoa*, it is fitting to add that *no story is landlocked*.

Biblical interpretation is, in many ways, landlocked. The ways and experiences of islanders, and the various notions of islandedness (some of which are discussed in this collection of essays), can help open up the landlockedness of biblical interpretation. In inviting and welcoming the rippling of *talanoa* (conversation), i reiterate that every story is an island even if not all readers are aware of their (stories and readers) islandedness.

Works Cited

Alter, Robert. 2004. *The Five Books of Moses: A Translation with Commentary.* New York: Norton.
Brenner, Athalya, ed. 1999. *Ruth and Esther.* FCB 2/3. Sheffield: Sheffield Academic Press.
Bush, Frederic W. 1996. *Ruth, Esther.* WBC. Dallas: Word.
Eskenazi, Tamara Cohn, and Tikva Frymer-Kensky. 2011. *Ruth.* JPS Bible Commentary. Philadelphia: Jewish Publication Society of America.
Fewell, Danna N., ed. 1992. *Reading between Texts: Intertextuality and the Hebrew Bible.* Louisville: Westminster John Knox.

---

7. Used in the Tongan Methodist translation of the hymn "Rock of Ages."

Fewell, Danna N., and David M. Gunn. 1993a. *Gender, Power, and Promise: The Subject of the Bible's First Story.* Nashville: Abingdon.

———.1993b. *Narrative in the Hebrew Bible.* Oxford: Oxford University Press.

———.1999. "'A Son Is Born to Naomi!' Literary Allusions and Interpretation in the Book of Ruth." Pages 232–39 in *Women in the Hebrew Bible.* Edited by Alice Bach. New York: Routledge.

Figiel, Sia. 1999. *They Who Do Not Grieve.* Auckland, NZ: Vintage.

Greenstein, Edward L. 1999. "Reading Strategies and the Story of Ruth." Pages 211–31 in *Women in the Hebrew Bible.* Edited by Alice Bach. New York: Routledge.

Havea, Jione. 1995. "The Future Stands between Here and There: Towards Islandic Hermeneutics." *The Pacific Journal of Theology* 2/13:61–68.

———.1996. "Shifting the Boundaries: *House of God* and the Politics of Reading. *The Pacific Journal of Theology* 2/16:55–71.

———. 1998. "*Tau lave!*" [Let's Talk]. *The Pacific Journal of Theology* 2/20:63–73.

———. 2003. *Elusions of Control: Biblical Law on the Words of Women.* Atlanta: Society of Biblical Literature; Leiden: Brill.

———. 2008. "'*Unuʻunu ki he loloto,* Shuffle Over into the Deep, into Island-Spaced Reading." Pages 88–97 in *Still at the Margins: Biblical Scholarship Fifteen Years after Voices from the Margin.* Edited by R. S. Sugirtharajah. New York: T&T Clark.

———. 2009. "Local Lectionary Sites." Pages 117–28 in *Christian Worship in Australia: Inculturating the Liturgical Tradition.* Edited by Anita Monro and Stephen Burns. Strathfield, NSW: St. Pauls.

———, ed. 2010. *Talanoa Ripples: Across Borders, Cultures, Disciplines.* Albany, NZ: Pasifika@Massey University Press; Auckland, NZ: Masilamea.

———. 2015. "Routes/Roots in Oceania: Migration, in Sea Views." Forthcoming in *Migration and Church in World Christianity.* Edited by Elaine Padilla and Peter Phan. New York: Palgrave.

Honig, Bonnie. 1999. "Ruth, the Model Emigrée: Mourning and the Symbolic Politics of Immigration." Pages 50–74 in *Ruth and Esther.* A Feminist Companion to the Bible 2/3. Edited by Athalya Brenner. Sheffield: Sheffield Academic Press.

Hubbard, Anthony, and Nicky Hager. 2012. "WikiLeaks Proves Brutal US Diplomacy." Stuff.co.nz. http://www.stuff.co.nz/national/politics/6519429/WikiLeaks-proves-brutal-US-diplomacy.

Jeansonne, Sharon Pace. 1990. *The Women of Genesis: From Sarah to Potiphar's Wife*. Minneapolis: Fortress.
Kaltner, John. 2003. *Inquiring of Joseph: Getting to Know a Biblical Character through the Qur'an*. Collegeville: Litugical.
LaCocque, André. 2004. *Ruth: A Continental Commentary*. Translated by K. C. Hanson. Minneapolis: Fortress.
Linafelt, Tod, and Timothy K. Beal. 1999. *Ruth and Esther*. Berit Olam. Collegeville: Liturgical Press.
McGee, J. Vernon. 1988. *Ruth and Esther: Women of Faith*. Nashville: Nelson.
Sarna, Nahum M. 1989. *Genesis*. JPS Bible Commentary. Philadelphia: Jewish Publication Society of America.
Scullion, John J. 1986. *Genesis 37–50: A Commentary*. Minneapolis: Augsburg.
Spivak, Gayatri Chakravorty. 2008. *Other Asias*. Malden, MA: Blackwell.
Vaka'uta, Nāsili, ed. 2011. *Talanoa Rhythms: Voices from Oceania*. Albany, NZ: Pasifika@Massey University Press; Auckland, NZ: Masilamea.

Second Waves

Sand, Surf, and Scriptures

Roland Boer

There is something distinctly new about this collection of essays—a rare experience in the vast production of knowledge that now happens under what counts as scholarship. It is perhaps one of the most enjoyable and thought-provoking edited volumes I have encountered for some time in biblical criticism. My task is to respond to three of the chapters, those by Steed Vernyl Davidson, Nāsili Vaka'uta, and J. Richard Middleton, as well as the introduction written by the three editors. My response takes each in turn, before seeking some common themes and problems. In doing so, I begin with a story and then focus on a key question that arises from each, a question that emerges from various other features of the argument.

Island Ambivalences

As I read "RumInations," the joint introduction, a story returned to me. It was triggered by the inclusion of Australia as a "roomy island" (2), along with Iceland, Madagascar, Papua, Solomon, and Aotearoa. At school we were taught repeatedly that Australia is unique, for it is an island continent, the only such landform in the world. How special, we thought, a claim to being different and perhaps better than anyone else. We may live on the smallest continent in the world, but we are also the largest island. I remained secure in this knowledge for some three decades only to have my assumptions concerning Australia turned on their head. Why? I found out at last that an island cannot be a continent and vice versa. An island must have the effects of the sea, weather-wise, felt across its interior. On this matter, Australia does not fit the bill. Anyone who has been in the vast deserts of the interior can certainly attest that the sea is far, far away and that the sea's effects are long gone. Australia is the smallest continent with no claim to being the largest in respect to anything. So what is the largest island? Greenland, half a world away and a little cooler.

I am less interested here in asking the question, What is an island? Others will focus on that question. Instead, the core question I have to ask of "RumInations" is somewhat different: Is the colonial narrative one of the relations between continent and island, or is it also between islands and islands? Much of the introduction operates with the first opposition. Colonizing endeavors set out from continents (at least inhabited ones), and they subdue islands—have done and continue to do so. This enables a particular construction of islands, thereby producing a range of insights. For example, islands are not impenetrable fortresses, but by their nature they are permeable and interstitial. The key relationship is not dialectical but trialectical, for it includes the sea. Indeed, the ever-present nature of the sea, its winds and waves and storms and calms and currents, characterizes both the immediate reality of islands and the relations between then. They are really periods in the sea. So, in perhaps the best sentence of the introduction, we find "As people who live in water do not know what it means to be wet, islanders who are isolated from everybody else do not see isolation as a problem" (17). I read this not only as an observation on the mutual constitution of islands and the sea in relation to one another, not only as a parallelism that draws out the analogy, but also as a dialectical observation. Isolation can only be understood through the connectivity provided by the sea, so much so that it redefines the definition of isolation.

To return to my question: what about islands and islands, especially when the connectivity of the sea becomes a means of colonial domination by one island over another?[1] Imperial islands are more common than we care to think. Those islands off the western peninsula of the Eurasian landmass are but one example, but others include Denmark, Venice, Japan, and Tyre. The chapter by Hisako Kinukawa deals with the last two on my list, and I know that Andrew Mein focuses on the United Kingdom in his response. So I would like to tack differently and steer towards the editors' intriguing insight that the volume does not deal extensively with the negative connotations of islands (12). Although I can see the importance of valorizing an island hermeneutics that challenges their characterization

---

1. What about river islands? These are often defensive locations from medieval times, where the river flowing around an island creates a natural barrier. Paris and Montreal come to mind, as does New York. At other times, they become industrial locations, as in Newcastle, Australia.

as isolated, remote, backward, and uncivilized, I am intrigued by the types of islands mentioned here. They include infamous prison islands such as Robben, Elba, and Alcatraz, but also "barrier islands" used by continental powers as defense lines and buffer zones. Here too are ambivalent islands appropriated and exoticized like Hawaii and the French Antilles and those vilified like Cuba and Taiwan. I would add those that become part of the contest between powers (whether continental or island based) over lucrative fishing and mineral zones. The uninhabited Diaoyu/Senkaku Islands are currently the focus in a renewed struggle between China and Japan, as in the east Timor Sea as East Timor challenges Australia's earlier shifty deal to gain access to its waters. Indeed, this feature is as old as islands themselves. As far back as empires go one of their key objectives has been to seize and control strategically positioned islands for all manner of reasons. The brief paragraph on these negative connotations closes with a couple of questions: "As we continue thinking together about islands and biblical interpretation, what might we learn by taking seriously these islands also? How might they contribute to a broader understanding of 'island' reading?" (12).

I look forward to the efforts to answer those questions as the debate continues. One item that may be worth considering actually comes from Antonio Negri, although it was first formulated by Lenin. In his effort to rethink the whole shape of radical political thought, Negri stopped writing and worked intensely with the industrial unions on the Italian coast off Venice (another island power for a time). In the collectives among whom he worked, they began to develop a theory of *operaismo*, or workerism.[2] The nub of that theory was the insight that resistance does not take place

---

2. It was developed in two journals, *Red Notebooks* (*Quaderni Rossi*, 1961–1965) and *Working Class* (*Classe Operaia*, 1963–1966). One of the best definitions is as follows: "*Operaismo* builds on Marx's claim that capital reacts to the struggles of the working class; the working class is active and capital reactive. Technological development: Where there are strikes, machines will follow. 'It would be possible to write a whole history of the inventions made since 1830 for the sole purpose of providing capital with weapons against working-class revolt.' (*Capital*, Vol. 1, Chapter 15, Section 5) Political development: The factory legislation in England was a response to the working class struggle over the length of the working day. 'Their formulation, official recognition and proclamation by the State were the result of a long class struggle.' (*Capital*, Vol. 1, Chapter 10, Section 6) *Operaismo* takes this as its fundamental axiom: the struggles of the working class *precede* and *prefigure* the successive re-structurations of capital" (Negri and Hardt 2002).

in response to oppressive state power but is constitutive of that power. In other words, the real driving force of history is precisely that constitutive resistance to which oppressive state and imperial powers must constantly adapt, must find new ways of trying to contain. The implications of this position are obvious for Negri's communist project, but I would also like to suggest that dominated islands too may be seen in this light. It is not for nothing that imperial powers have sought to dominate and control islands, from Alexander the Great over Tyre, through the Caribbean and the spice islands of the East Indies, to Guam and American Samoa today. They provide continued forms of constitutive resistance that ultimately cannot be dominated.

Biblical Thalassophobia?

"Building on Sand," the title of the chapter by Davidson, triggers another story. It is based on the parable of the wise and foolish builders from Matt 7:24–27.[3] Here the wise man builds his house upon the rock, where it withstands rain, flood, and wind. This, suggests Matthew's Jesus, is like the one who hears his word and acts upon it. But not the man who builds his house upon the sand: it too suffers rain, flood, and wind, but in this case the house on the sand fell flat. Of course, the last phrase is not from Matthew's Gospel ("it fell—and great was its fall" reads that text), but from a children's song based on the parable. I spare the reader the song itself, but it reinforced in me the belief that houses on rocks are solid and that houses on sand are unstable, prone to fall whenever the weather turns nasty. Imagine my surprise when (in my twenties) I was discussing such matters with a builder: he pointed out to me that the best foundations for a house are placed not on rock but on sand. The biblical parable has it all wrong. Why? Rock is inflexible, and it suffers from extremes of hot and cold. It expands and contracts, cracking when movement is too great. Into a crack water gathers, and, if it freezes and thereby expands, it makes the crack worse. By contrast, sand is flexible and manages temperature extremes very well.

---

3. "Everyone then who hears these words of mine and acts on them will be like a wise man who built his house on rock. The rain fell, the floods came, and the winds blew and beat on that house, but it did not fall, because it had been founded on rock. And everyone who hears these words of mine and does not act on them will be like a foolish man who built his house on sand. The rain fell, and the floods came, and the winds blew and beat against that house, and it fell—and great was its fall!" (NRSV)

Water does not gather, but seeps down through the sand. And when you place a weight on sand, it is extremely firm.

So building on sand is far better, which is good news for Davidson's study. He focuses on that in-between zone between land and sea, the sand of the beach upon which the tide flows in and out. Sand is neither land nor water (dig down in the sand, and soon you come upon water). Indeed, Davidson identifies four features of sand that shape his biblical analysis (of Gen 38 and Dan 8): shiftiness, doubled quality, motion, and insularity. The first offers a welcome criticism of Homi Bhabha's hybridity, as also of the colonizing agenda of contextualization. In its place, Davidson proposes a process of revision and re-creation, for sand is constantly on the move, creating new "shapes, lines, sizes, and textures out of the old shoreline" (40). So also with the Bible, a text that has washed up on the shoreline of islands. The second concerns the doubled quality of sand, troubling a whole series of oppositions (like the limbo dance)—beginning and end, old and new, oppression and freedom, oppression and self-determination, land and sea. Here Davidson engages briefly with Gen 38, exploring the complexities of doubleness between dead sons and new sons, between Tamar as wife and mother of brothers, and thereby between the doubled nature of levirate marriage. The same text appears in the treatment of the third hermeneutical item, namely, tides and rhythms. Now time itself is reconfigured, for tidal patterns produce time not in terms of definitive progression that is marked in its passing. In Gen 38 this becomes the melding of actions, the interplay of events and nondescript periods: "again," "yet again," "at that time," pepper the story, alongside times of mourning and of delivery. The final hermeneutical point asks how sand influences notions of insularity. The effect is quite momentous, for emancipation becomes not a process of apocalyptic liberation, but rather a noncataclysmic process that draws upon the shifty nature of sand, its doubled quality, and tidal time. Obviously, this final category opens up alternative readings of apocalyptic texts in the Bible, and Davidson's focus is Dan 8.

I have not as yet asked my main question concerning Davidson's proposal: what does one do with the significant feature of biblical thalassophobia? In this respect, the economically and politically marginal zone of the southern Levant (in which Israel appeared late in the story) shares a common feature of ancient southwest Asia. The sea was an absolute limit to imperial expansion, a boundary marker that was threatening and fear-engendering. Myths such as *Enuma Elish* depict the sea as female, destructive, and needing to be overcome. In the Bible, the flood and the Red Sea are by no means benign forces; Jonah suffers when he sets sail; and Paul too comes to grief when he takes to the sea. Overall, references to the sea are few and far between. What about Tyre, Sidon, and Byblos? The fact that these towns are condemned in the Bible (for example, Ezek 27–28) and in the realm of imperial politics were left to largely their own devices, instead of being easily appropriated into the shaky realms of aspiring potentates, only reinforces such a thalassaphobia (Revere 1957, 40). The question is how one deals with such a literary and geopolitical fear of the sea, especially in the search for an islander hermeneutics in which sea, sand, and tides are crucial.

We may develop deeper modes of analysis, as Davidson does with the categories of shiftiness, doubleness, rhythm, and insularity, but I wonder whether his interpretation of Dan 8 may offer some guidelines for engaging with thalassophobia itself. He describes this text as both engaging and distancing. It is distancing for the great struggle between the flawed proponents, the ram and the goat, since that seems pointless for island readers. Is that the case too with thalassophobia? Instead, he proposes that we should focus on the neglected animals that remain, marked by "all beasts." These are, after all, the majority, so how do they fare? How do they construct their subjectivity outside the pointless struggle? Or, in terms of thalassophobia, are there items that slip out of its purview, that have thus far been neglected in interpretation?

An Oceanic Manifesto

The chapter by Vaka'uta reads much like a manifesto. Clear formulations and sharp observations are undergirded by a persistent and welcome militancy. I will return to text of that manifesto in a moment, but first, a story. It is triggered by the definition of what Vaka'uta calls island-marking: it "refers to a reading of the Bible that *arises out of island contexts, shaped by island cultures and values, gives privilege to island knowledge-systems (epistemologies) and languages, reads the Bible through island/oceanic lenses, takes account of critical issues that confront islanders, and serves the interests of the islands and islanders*" (57, emphasis original). The focus is clearly Oceanic, those islands that are constituted and defined by the Pacific Ocean.

To the story: A couple of years ago, I was on a voyage across the Pacific on a container ship, a sixteen-day crossing from Aotearoa to Panama as part of a longer voyage halfway around the world. The glory of a voyage like this is that no "entertainment" is provided, so you are free to spend time on deck or in the bridge, read, write, and watch old DVDs in the common room. One of those DVDs was what might be called a prequel to the Indiana Jones series of films. "The Treasure of the Peacock's Eye" it was called (Selbo 1995), and it was truly a B-grade movie (not a faux B-grade like the other Indiana Jones films). Woeful acting, unbelievable plot lines, dreadful filming and editing. It told the story of a young Indiana Jones and his first adventure, which was to seek out some archaeological treasure called the Eye of Peacock. This search took him in swashbuckling fashion across Europe, the Middle East, Asia, and then the Pacific Islands. More specifically, he was shipwrecked, along with his overweight partner in arms, on one of the Trobriand Islands. Here he met none other than Bronislaw Malinowski, who turned out to be a countercultural type *avant la lettre*, who had turned his back on Western "civilization" and was learning to live once again from the islanders. After lengthy conversations with Malinowski, the young Indiana Jones realizes the error of his ways and vows to give up the vain pursuit of archaeological treasures. Obviously, this prequel negates the very premise of the three other Indiana Jones movies, thereby rendering them impossible.

I tell this story in part because Malinowski was, of course, the first theorist of the economic categories of reciprocity and redistribution. These he learned from his study of the Trobrianders, and they were to make their way into economic studies of ancient southwest Asia and biblical societies through the work of Karl Polanyi. But I also detect a significant theme of what Malinowski called reciprocity and redistribution in the chapter by Vaka'uta. As the term suggests, reciprocity designates the socially determined economic relations of mutuality, of the give and take of symmetrical social arrangements. Redistribution depends on a level of centralization, in which goods are gathered by a central administration and then distributed to where they are needed (Polanyi 2001, 49–55; Malinowski 1932). Combined, reciprocity and redistribution are the two planks of exceedingly complex social and economic systems, in which exchange takes place without the profit motive.

For Vaka'uta the key term is *fale-'o-kāinga*, which he defines as a combination of the subtle semantic fields of *fale* (the many senses of household) and *kāinga*. The second has a base sense of kinship, but is actually a cultural network of relationship and exchange. Crucially, he describes the core value of *kāinga* as reciprocity. In Tongan, this is *tauhi vā*, while in Samoan it is *teu le vā*: "*Tauhi vā* elevates distribution above consumption, sharing above accumulation, peaceful coexistence above domination, communal well-being above individualistic interests" (60). By now, the connection with my story should be obvious, although I am the last one to insist that Malinowski did it first or described it best (even though no criticism of Malinowski appears in Vaka'uta's chapter).

What has all this got to do with biblical criticism? Vaka'uta provides six features of *fale-'o-kāinga* that then become the basis for guidelines for biblical criticism: undeserved respect for one another and the earth (all embodied in *fale*); wisdom in managing resources; sharing what one has no matter how little; fulfilling duties and obligations to those in the *kāinga* network; sincerity and devotion in sharing; and showing humility. Translated into the practices of biblical criticism, which takes place within a community that includes the whole of nature and involves a network of relations that knows no borders, we are urged to (1) show respect to biblical characters and readers, especially those marginalized because of sexual orientation, ethnicity, social position, and religious beliefs; (2) read the Bible wisely rather than aggressively, aware that academic sophistication blocks out many readers; (3) participate in the process of interpretation, in a way that challenges those who feel they have the correct answers; (4) read responsibly, especially with an awareness of the purpose of interpretation; and (5) be humble, since one is all too aware of the limitations of one's biblical interpretation. In sum, within an interdependent community, Vaka'uta urges us to be respectful, wise, participatory, responsible, and humble.

In closing, I would like to pose a question: while I affirm each of the points of Vaka'uta's manifesto, how does one deal with the many biblical stories that are not respectful, wise, participatory, responsible, and humble? These stories are legion in the Bible, from the crushing, in the name of God, of perfectly justifiable rebellions to the unaccountable sexual abuse and brutality. How does one read them?

Reforming Preachers

Like the chapter by Vaka'uta, the one by Middleton is also from a committed, faith-based perspective. That leads me to my story, which emerges from a recent experience of editing a large volume with Fernando Segovia. As we gathered material for *The Future of the Biblical Past* (Boer and Segovia 2013), we found that those who practice biblical criticism in an "academic" and "scientific" manner are clearly in the global minority. The vast majority of those who read and interpret the Bible do so from within what may be called a faith community (I deliberately leave the term faith undefined). What about that minority? They study the Bible in a "purely" academic context, assuming that matters of faith and community hamper "scientific" inquiry. That this is a specialist approach, borne out of the needs of the growth of academic "disciplines," should be clear. Indeed, it is a feature of all academic disciplines as they sought a verifiable status within universities (Wallerstein 2011, 264). Yet this approach by the minority has come to be the dominant hegemony in the study of the Bible.

Middleton speaks unashamedly from and to the majority, not the minority, and thereby challenges the hegemony of academic biblical criticism. He seeks to reflect upon and contribute to the way everyday Caribbean Christians read the Bible, challenges the dominant theological approaches to salvation within the Caribbean churches, explores the way hymns reflect that theological position, and then closes with three suggestions for the reform of preaching from the Bible in the churches. The core of his argument is to reshape the way salvation is understood. Against a view of salvation as other-worldly, taking place in the context of a romanticized nature, he proposes a salvation that is this-worldly and natural.

That is, he seeks a biblical Caribbean theology that grounds human liberation in God's intent for creation. In doing so, he criticizes European, North American, and Caribbean biblical interpretation (which has been far too dependent on the other two) for neglecting the presence of nature within biblical texts. For this reading, he engages a full range of texts from the Hebrew Bible and New Testament in order to argue that salvation concerns the restoration of the earth and is not merely focused on the heavens above. In a comparable fashion, he contrasts the hymnody of the Caribbean churches—in which "Christians don't tell lies; they just go to church and sing them" (Middleton in this volume, 128, quoting A. W. Tozer)—with Caribbean music and its affirmation of nature. In a wonderful twist, the Bible thereby lines up with that music and not that of the church. Finally, he offers three proposals for reforming Caribbean churches through a biblical creation theology: it will provide (1) a foundation for the holiness of everyday life lived in terms of the *imago Dei* in society, (2) an ethical challenge to the unjust status quo, and (3) an energizing of church members as they work to improve a world in which salvation is being worked out.

In this case, my question is quite simple: While I can appreciate the need to draw upon and fix one's focus on affirming a world undergoing salvation, what about those texts that do deny the earth and our material existence? Do these texts become the ones that are cited as negative examples, as ones to avoid? Do they provide a focus of resistance, against which one argues for a biblical theology of natural salvation? Are they seen as aberrations of the core message of the Bible, or are they part of its constitutive tension? Or does one, after having noted and argued with such a tension in the biblical material, take sides for the earth? My questions have multiplied, but they arise from Middleton's argument in an urgent manner. If one is to engage with faith communities, then these too need to be part of the debate and struggle.

Conclusion

Four stories and four questions, in what is a provocative and thoughtful collection of chapters: the role of negatively coded islands and resistance; the problem of biblical thalassophobia and island readings; the challenge of texts that are not respectful, wise, participatory, responsible, and humble; and the ways one engages with texts that counter a biblical theology of this-worldly and natural salvation. I ask these questions not by way of knocking down any of the arguments put forward in these chapters, but as a way of continuing the dialogue.

Works Cited

Boer, Roland, and Fernando Segovia. 2013. *The Future of the Biblical Past: Envisioning Biblical Studies on a Global Key*. Semeia Studies. Atlanta: Society of Biblical Literature.

Malinowski, Bronislaw. 1932. *Argonauts of the Western Pacific: An Account of Native Enterprise and Adventure in the Archipelagoes of Melanesian New Guinea*. London: Routledge.

Negri, Antonio, and Michael Hardt. 2002. "Marx's Mole is Dead! Globalisation and Communication." *Eurozine* 13. http://www.eurozine.com/articles/2002-02-13-hardtnegri-en.html.

Polanyi, Karl. 2001. *The Great Transformation: The Political and Economic Origins of Our Time*. Boston: Beacon.

Revere, Robert B. 1957. "'No Man's Coast': Ports of Trade in the Eastern Mediterranean." Pages 38–63 in *Trade and Market in the Early Empires: Economies in History and Theory*. Edited by Karl Polanyi, Conrad M. Arensberg, and Harry W. Pearson. New York: Free Press.

Selbo, Jule. 1995. *Young Indiana Jones and the Treasure of the Peacock's Eye.* Directed by Carl Schultz. Universal City, CA: Amblin Television; San Francisco, CA: Lucasfilm; Hollywood, CA: Paramount Television.

Wallerstein, Immanuel. 2011. *The Modern World-System IV: Centrist Liberalism Triumphant, 1789–1914.* Berkeley: University of California Press.

ISLANDEDNESS, TRANSLATION, AND CREOLIZATION

Aliou C. Niang

Island, Islanders, and Bible: RumInations is a collection of engaging essays inviting readers to join in "a conversation on how being islanders, and the various ruminations of islandedness, condition the way" Scripture is read (Davidson, Aymer, and Havea in this volume, 2). Place and identity not only shape the reading of and thinking through biblical texts (12–19), they are integral to reimagining the "hermeneutics thing" (20) in conversation with inlanders and nations (28–30). Therein lies the invaluable contribution *RumInations* makes to reading and interpreting biblical texts contextually and yet in conversation with other interpreters.

This essay engages those by Margaret Aymer, Mosese Ma'ilo, and Althea Spencer Miller. Each of these essays reflects striking personal experiences of a postcolonial condition that resonates with me as a native of Senegal, West Africa.

I grew up in Adéane, a village about forty-four miles inland from the island of Carabane in the Casamance region once occupied by Emmanuel Bertrand-Bocandé (1812–1881), a French colonial agent steeped in implementing French culture in the region (see Bertrand-Bocandé, Debien, and Martin 1969, 279–302; Bertrand-Bocandé 1856, 398–418). I later moved to Dakar where I often visited the island of Gorée, famous for its *Maison des Esclaves* ("House of Slaves") two miles east of the port of Dakar on the Atlantic Ocean. Gorée's strategic location made it a much coveted place by European slave traders and later colonialists. In 1444, the Portuguese, under the leadership of Dinis Diaz, sailed to the island on which they later built trading posts. The attractiveness of its location generated fierce conflicts between European traders. The Dutch seized it from the Portuguese and by 1588 renamed it Goede Reede ("Good Port"), a name that evolved into Gorée. Next came the British who displaced the Dutch and occupied the island from 1664 to 1667 but were finally forced out by the French, who controlled it from 1677 through much of the colonial era (Niang 2005, 320). Today, Gorée is visited by numerous Europeans and Americans not as much for its colonial history but its museum, the *Maison des Esclaves*, which revisionists now question whether it actually played any significant role in the slave trade (Mack 2011, 40–44). In spite of these critiques, Gorée did play a role, and downplaying its importance is to ignore the dehumanizing nature of the transatlantic slave trade itself.

Gorée was significant in the Atlantic Slave Trade and French colonization of Senegal. It was one of the *Quatre Communes* (the "Four Towns")[1] where the assimilation policy was first implemented in Senegal in French colonial West Africa from 1887 to 1960. French citizenship was granted to dwellers of the "Four Towns" known as "*originaires*" and their progeny with rights metropolitan French people enjoyed while the rest of the Senegalese lived under a protectorate. The same applies to the inhabitants of the island of Saint-Louis, about 163 miles north of Gorée, which in turn is about 143 miles from Carabane by boat. Echoing Aymer's Pauline "islands of accommodation," these "towns" are islands of assimilation. I will now summarize the main ideas of each essay with a concise response.

Islandedness, Paul, and John of Patmos

Two poems from two islanders, Robert Nesta Marley and Louise Bennett-Coverly, spearhead Aymer's intriguing work—epigraphs that enshrine an antithesis she detects in the Pauline greeting "grace and peace" and Johannine exhortation to "come out of Babylon." To delve into what might have given rise to these oppositional perspectives, Aymer probes the idea of "islandedness and how it might function as a wedge with which the study of the Bible and the ancient world might be split open in new ways" (in this volume, 35). She coins and uses the word "islandedness" as a thinking metaphor with which to understand the relationship between "phenomena: roots and routes" (36–37). The former, she avers, deals with identity construction and the latter with various forms of displacements (migrations). What follows is a sustained and insightful observation of how dis-

---

1. See Sembène 1981, 134–35. The other three towns include Saint-Louis, Dakar, and Rufisque.

placement "routes" affect identity "roots" and the inescapable role of the sea in the making. She argues that "diaspora space" instead of "insularity" is an identity shaper, and the latter is used by the islanders "to establish and express social identity" in response to islandedness (39).

Aymer rightly sees assimilation, marginalization, accommodation, and alienation as variable modes in which "diaspora space" is negotiated depending on what is required for adjustment in the host culture. The poems of Marley and Bennett-Coverly offer good examples of this. Whereas Marley rejects the culture that shaped him (42), Bennett-Coverly accommodates to the same "diaspora space" with her Jamaican Patois as an identity constructor. Aymer sees this phenomenon operative in Pauline and Johannine diaspora space negotiations in antiquity. According to Aymer, John rejects accommodation for alienation whereas Paul embraces a sort of "early church Patois" anchored by his "grace and peace" salutation that welcomes all people (49). Further, Paul is read as one who navigates seas to create islands of accommodation, say house-churches, whereas John is so ambivalent about the sea that he envisions a new heaven and earth with no sea (Rev 21:1). Aymer closes with probing questions for individual and corporate islanders and inlanders to ponder.

To Aymer's point, Scripture provides clues into how our ancestors in the faith worked through their diaspora-spaces, especially evident in her discussion of Paul and John. As a voluntary immigrant myself, who now lives on the island of Manhattan, I find Aymer's understanding of Marley and Bennett-Coverly intriguing and her reading of Paul and John fascinating.

I wonder, however, what her conclusions would have been if she were to engage Gregory Stephens's insightful work entitled *On Racial Frontiers*, which emphasizes Marley's rejection of and desire to do away with Babylon's alienating presence (Stephens 1999, 148–220; see also MacNeal 2013). I wish she discussed more texts from both Paul and John that embed common themes. For instance, beyond "grace and peace," John's apocalyptic language somewhat echoes Paul's. Paul metaphorically talks about the transformation of the believer's life (living or dead) wrought by Christ's return (1 Thess 4:13–18)—an event the earth (creation) also awaits (Rom 8:19–23), which would include the sea. If John speaks of the fall of Babylon, Paul implicitly expected it. Paul and his Philippian converts belong to a heavenly citizenship/commonwealth instead of Judea or Diaspora (Phil 3:20) where he spent most of his life (Holladay 2003, 429–60). In spite of these minor critiques, Aymer's point is well taken.

Celebrating Hybridity

Ma'ilo documents how "the experimental translation of island Bibles in the nineteenth century was controlled by Western missionaries (esp. London Missionary Society and Wesleyan Missionaries)" (in this volume, 105). Since linguistic limitations influenced the cultural politics of difference, the resulting Bibles were affected by the translators' colonial impulses to "convert, dominate, and redirect the 'savage' islanders' moral, spiritual, and cultural consciousness through the power of language, *island Bible languages*" (107, emphasis original). Ma'ilo sees a need for liberating translation. Building on Mehrez and Bhabha, Ma'ilo thinks a hybridized Bible translation would enliven islanders and allow God's word to be heard (110)—an appropriate Samoan Bible translation that does not settle for simplicity.

Ma'ilo provides helpful examples of his emancipatory translation by showing how Samoan Bible translators settled for simplicity rather than appropriateness. For instance, they translate Matthew's "'with child' into *to*, which is the simplest and greenest term for pregnancy. They ignored other appropriate and more respectful terms in the Samoan idiom such as *tau'ave le tama*, meaning 'to bear about' … or 'to carry,' which best serves Matthew's 'with child'" (114). In the same vein, the most appropriate term for Mary's pregnancy in Samoan—*toifale* or *tofale* ("pregnancy outside of marriage")—is avoided for *to*, which also "confuses the poetics of the passage" and "leaves the situation opened for Samoan readers to be branded as *tofale* or otherwise" (115). Ma'ilo finds the rendition of the term virgin in Matt 1:23 as *taupou* problematic, because the term denotes "status and responsibility" and may not have anything to do with virginity in Samoan (116). On Jesus's identity as son of Mary in Matt 1:18–25, Samoans would have called him *tamaalepō* ("child of the dark")—a symbolic Samoan expression for a person born out of wedlock expected to suffer shame.

Finally, Ma'ilo turns to his hermeneutics of liberation to read *tamaalepō* ("child of the dark") with *tamaaāiga* ("person of large family connections") (118). Jesus's extensive genealogy (Matt 1:1–14) did not exempt him from social oppression, because he was a "child of the dark." According to Ma'ilo, Jesus's followers reinterpreted his teachings and death by reimagining him from a "child of the dark" to a person whose lineage reaches back to Abraham. This liberating reading of Scripture, Ma'ilo insists, is what islanders need to formulate.

I concur with Ma'ilo's call for an appropriate translation and liberating hermeneutics. Alternative translations he proposed are convincing. Although some missionaries try to learn the culture, create an alphabet, and then translate either the New Testament or the entire Bible, the temptation to regard their translations as normative undercuts their efforts. The proliferation of Bible translations in America should have been a constant reminder that cultures need translations that are linguistically appropriate and hermeneutically liberating. My point is this, "God woos us in tones and accents that belong with our primal self-understanding," because "Christianity is recognizable only in the embodied idioms and values of the cultures in which we find it, allowing Christians to speak and respond with the facility of the mother tongue and mother-tongue speakers with faith and trust in God's promise" (Sanneh 2012, 35–36). Ousmane Sembène, a Senegalese filmmaker, once posited that he wrote his movie scripts for one of his countercolonial movies entitled *Emitay* ("God") in Diola language instead of French to capture cultural linguistic appropriateness. To have Diola actors (one of the Senegalese ethnic groups) speak "proper" French, he insists, "would have been inappropriate" (Sembène 1993, 1–7). Ma'ilo's option for hybrid Bibles whose language "occupies ... Third Space" is clear (107–9, 116), but it would have been helpful if he were to clearly distinguish bilingualism from linguistic hybridity. However, as one who speaks many languages, I am unsure what a hybrid Bible translation would look like.

Creolizing Hermeneutics

Miller draws from the pioneers of the Negritude movement with special attention to Glissant's work. The work explores crucial dimensions of "Glissant's thought in conjunction with a sense of Caribbean islandedness" (in this volume, 126), that is, a "Caribbean lived experience" that underpins the entire work, guides, and moves its delicate community-centered and open-ended hermeneutic (127). A "transnational biblical conversation" (127) spearheads her quest for a liberating interpretive direction that begins with panoramic snapshots of the author's journey from the familiar to unfamiliar past, unearthing powerful memories of connections of a lost African ancestry through music and dance at a Methodist church service in Saint Thomas, Jamaica.

To Miller, the church service enshrined cultural encounters (wrought by colonization) that gave rise to her creolized identity (130). Drawing from Glissant, her creolization is a continuous process that "ensures the failure of totalizing, systematized, linear progressive, Eurocentric History" (133). It destabilizes colonial hegemony in its making a way from assimilation to a "radical inclusion ... reenculturation" (138). As she engages the conclusion of the Enlightenment, especially Hegelian dismissive comments on Africa and Africans (and Aimé Césaire's role and others in rebutting it), Miller returns to Glissant's main thought—his disinterest in retrieving the past and option for "reversion."

Strikingly, the voice of the subaltern *Jamaican Patoi* in speech or song, in Miller's estimation, is an integral aspect of creolization in the sacred sphere (144–45). Creolization is a dynamic agent that makes "reversion and recuperation" actionable. She then rightly makes a fine distinction between Glissant's and the postcolonial contributions of Edward W. Said, Homi K. Bhabha, and Gayatri C. Spivak while acknowledging some point of connection. She highlights how Glisaant's work is informed by his deep experience of a people doubly alienated by being displaced from their African history who strove to create a creolized culture. The author wonders what the implication might be to read Scripture in such a dynamic context, say continuous journey (135). The ideas of Glissant and Leslie James proved crucial to this work but in the end Glissant's resounded: "reversion is the strategy that dismembers gaps and ruptures, shreds veils and re-members history disruptively" (150).

Miller's work is refreshing. I agree that creolization is not a model for retrieving a robbed past but a dynamic innovation that provides "a bottom-up view of culture, which allows popular elements of the native to reassert themselves against the top-bottom model implicit in romanisation" (Wallace-Hadrill 2008, 11). Since the "Roman world produces no creole languages" (13), I wonder, however, what biblical texts Miller, as an islander hermeneut, would have read in conversation with Glissant's thought. Here, Lamentations and some psalms come to mind. Glissant's (1989, 15–17) distinction between the religious experiences of diaspora Jews and those of the Martiniqans does not preclude an exploration of their common themes. My point is investigating common themes such as forced displacement, suffering, alienation, resistance, and innovation would have engendered fruitful conversations. A staged conversation between Léopold Sédar Senghor (1964, 82; 1977, 97–101) and Ousmane Sembène (1981, 134–35), who both envisioned a step forward instead of a mere return to precolonial African traditions, and Glissant would have been insightful.

Aymer, Ma'ilo, and Miller have offered helpful ways of engaging the biblical text contextually. The promise their work enshrines is the recognition that geography, experience, and language shape how one reads biblical texts. Their resounding message to readers of *RumInations* is this: *we, too, who were once silenced, have a sacred story to tell.*

Works Cited

Bertrand-Bocandé, Emmanuel. 1856. "Carabane et Sédhiou." *Revue Maritime et Coloniale* 2/16:398–418.

Bertrand-Bocandé, Jean, Gabriel Debien, and Yves Saint Martin. 1969. "Emmanuel Bertrand-Bocandé (1812–1881): Un Nantais en Casamance." *NTBIFAN* 31:279–302.

Glissant, Édouard. 1989. *Caribbean Discourse*. Translated by J. Michael Dash. Charlottesville, VA: University Press of Virginia.

Holladay, Carl R. 2003. "Paul and His Predecessors in the Diaspora: Some Reflections on Ethnic Identity in the Fragmentary Hellenistic Jewish Authors." Pages 429–60 in *Early Christianity and Classical Culture: Comparative Studies in Honor of Abraham J. Malherbe*. NovTSup. Edited by John T. Fitzgerald, Thomas H Olbricht, and L Michael White. New York: Brill.

Mack, Deborah L. 2011. "When the Evidence Changes: Scholarship, Memory, and Public Culture at the *Maison des Esclaves*, Gorée Island." *Exhibition*, Fall:40–44.

MacNeal, Dean. 2013. *The Bible and Bob Marley: Half the Story Has Never Been Told*. Eugene, OR: Cascade.

Niang, Aliou C. 2005. "Postcolonial Biblical Theology in Geographical Settings: The Case of Senegal." Pages 319–29 of *Reconstructing Old Testament Theology: After the Collapse of History*. Edited by Leo G. Perdue. Minneapolis: Fortress.

Sanneh, Lamin. 2012. "The Significance of the Translations Principle." Pages 35–49 in *Global Theology in Evangelical Perspective: Exploring the Contextual Nature of Theology and Mission*. Edited by Jeffery P. Greenman and Gene L. Green. Downers Grove, IL: InterVarsity Press.

Sembène, Ousmane. 1981. *The Last of the Empire: A Senegalese Novel*. Translated by Adrian Adams. London: Heinemann.

———. 1993. "Ousmane Sembène Responds to Questions for the Audience." *Contributions in Black Studies* 11: 1–7.

Senghor, Léopold Sédar. 1964. *On African Socialism*. Translated by Mercer Cook. New York: Frederick Praeger.

———. 1977. *Liberté III: Négritude et civilisation de l'universel*. Paris: Éditions du Seuil.

Stephens, Gregory. 1999. *On Racial Frontiers: The New Culture of Frederick Douglass, Ralph Ellison, and Bob Marley*. New York: Cambridge University Press.

Wallace-Hadrill, Andrew. 2008. *Rome's Cultural Revolution*. New York: Cambridge University Press.

The Wrong Kind of Island?
Notes from a "Scept'red Isle"

Andrew Mein

I am grateful for the invitation to respond to this collection of essays, which offers an effective demonstration of both the viability and the vitality of island hermeneutics. My initial reaction is a personal one, and it is to make the observation that I too am an islander. I was born and brought up in the British Isles, where the notion that we are an "island nation" has long been deeply ingrained in the national character.[1] Whether this particular island identity should be thought of as a good thing is a more complex question in the light of this volume. Although it brings together voices from a variety of island backgrounds, there is a strong tendency throughout the volume to define island experience over against Western, colonial interests and to avoid tracing the role the Bible played in that colonial enterprise, all of which leaves me wondering whether I might be the wrong kind of islander to make a contribution. Nevertheless, what I should like to do in the following pages is to relate three of the essays from the volume to a few

---

1. Shakespeare expresses the sentiment famously in *Richard II* (act 2, scene 1): "This royal throne of kings, this scept'red isle, / This earth of majesty, this seat of Mars, / This other Eden, demi-paradise, / This fortress built by Nature for herself / Against infection and the hand of war, / This happy breed of men, this little world, / This precious stone set in the silver sea, / Which serves it in the office of a wall / Or as a moat defensive to a house, / Against the envy of less happier lands,— / This blessed plot, this earth, this realm, this England."

In his popular study, *The English*, Jeremy Paxman (1999, 30) claims that "it would be hard to exaggerate the importance of the fact that they are islanders to the mentality of the English." Both he and Shakespeare speak of England and the English, who of course have to share the island of Great Britain with the Scots and the Welsh (not to mention the complex relationship between Britain and Ireland), but I believe that what Paxman says of the English here is also broadly true of a British identity.

aspects of British island history and to some of the ways in which the Bible has been used both to shore up that identity and to create the stereotypes which this volume so effectively challenges.

The proposition with which the whole volume opens is the twofold one that "biblical texts are like islands, and readers are like islanders" (Davidson, Aymer, Havea in this volume, 19). Jione Havea's essay "Sea-ing Ruth with Joseph's Mistress" brings both of these issues squarely within view as he works with two principal images: the Pacific Island practice of *talanoa* and the related idea of biblical stories as islands in a sea of telling. The practice of *talanoa*, a form of communication, storytelling, and conversation, without formal agenda or hierarchical structure, offers a model for connecting two biblical characters that are not usually spoken of in one breath—Ruth the Moabite and Potiphar's wife, or as Havea reasonably calls her, Joseph's mistress. In his reading he unmoors the stories from their biblical contexts and has them rub up against one another. This enables us to see common currents more clearly, such as the women's status as outsiders in the biblical world and as women in control of their own sexuality. It also accentuates differences—Ruth's need for stealth in fulfilling her desires over against the directness of Joseph's mistress when she makes the demand "Lie with me!"

Havea's *talanoa* not only emphasizes the importance and indeed inevitability of selectivity in biblical reading, but also grows out of his sense that these individual stories are themselves islands, not because they are isolated from one another, but because it is possible to travel directly from one to another. In this context he takes issue with the dictum, used from time to time in intertextual studies (see Miscall 1992 and ultimately adapted from John Donne's *Seventeeth Meditation*), that "no text is an island." He finds this highly problematic because of the way it romanticizes and isolates islands, seeing them as disconnected from one another by the barrier of the sea. Island experience is in fact the opposite of this: the sea is a means of connection and communication: "Distance is not barring, and home is a place of transit," and indeed "*every story, text, and talanoa is an island*" (Havea in this volume, 148–49, emphasis original). If the original sense of Donne's "no man is an island" is that human beings are fundamentally linked to one another, then the image fails because islands are themselves a crucial example of such connection.

Havea's images of the sea as a medium of connection are largely benign, but this need not be the case. The history of my own islands may serve as a reminder that a sea that enables communication can also enable conquest. In fact, Donne's characterization of the relationship between sea and island was not the only one operative in early modern Britain. The historian Jonathan Scott (2011, 14) comments that "in pre- and early modern Europe, when transport was only efficient upon water, seas did not divide, they connected." And Scott goes on to note that for early English travel writers like Samuel Purchas and John Evelyn, Shakespeare's dictum that the sea stood to England as a wall or a "moat defensive to a house" did not apply. Rather the sea was a bridge from the island to the rest of the world, over which both commerce and conquest were possible: the "Uniter by Traffique of al Nations" (Scott 2013, 30, quoting Samuel Purchas from 1625). It was thus precisely because of its easy access to the sea that my own island nation was able to extend its power and influence so widely around the globe.

One consequence of recognizing the darker side of such communication between islands is to question whether Havea's *talanoa* approach to Scripture might not also throw up more troubling and less playful juxtapositions. What happens if we read Ruth's sexual agency alongside the stories of Oholah and Oholibah in Ezekiel or the assertiveness of Joseph's mistress alongside Paul's encouragement for slaves to obey their masters? There is certainly scope here for an intriguing kind of antilectionary.

If Britain's island status provided the initial impetus toward its imperial ambitions, that was also true, if to a more geographically limited extent, for that other imperial island, Japan. Hisako Kinukawa begins her study of the Syro-Phoenician woman in Mark's Gospel with a brief account of Japan's role as an imperial island. Her recognition of the oppressive nature of Japan's relationship with its mainland neighbors provokes her to look for a similar context of imperial expansionism to make sense of this difficult story.

In Tyre Kinukawa finds an imperial island in the biblical world. Despite its tiny size, the city-state had a long history of expansion through trade and colonization, and even as part of Hellenistic and Roman empires it retained much of its wealth and power.[2] Kinukawa focuses on the his-

---

2. It is perhaps no surprise that Tyre and the Phoenicians provide one ancient model that appears quite frequently in nineteenth-century British self-understanding. Timothy Champion (2001, 456) describes how they were used as "prototypes of British imperial and commercial dominance." Amongst other examples, he quotes the ancient historian George Rawlinson's 1889 claim that the Phoenicians were "the people who of all antiquity had most in common with England and the English—the

torical situation of the first century CE and emphasizes the discrepancy between the affluent elite of the city of Tyre itself and the poor villagers of both the Phoenician and the Judean hinterland. In this context she follows Gerd Theissen in arguing that this story, set "in the region of Tyre," identifies the woman as part of Tyre's affluent elite (Kinukawa in this volume, 143; Theissen 1991). She is "a Greek, a Syro-Phoenician by birth" (Mark 7:26), which implies that she is part of the Greek-speaking elite, even if of local Phoenician stock. Jesus's anger at her request for help, and his reply in terms of table fellowship, are a reaction to one he saw as responsible for taking food out of the mouths of Galilean children.

It is striking that, on Kinukawa's reading, the imperial islander nevertheless has something to teach Jesus as well as something to gain from him. In this context, Kinukawa emphasizes the dual identity of the woman: on the one hand, she is to be identified with Tyre's ruling elite; on the other, she is marginalized as a woman and the mother of a demon-possessed child. Yet even this underplays the historical and cultural complexity of the woman's position. If Theissen (1991, 70–71), on whom Kinukawa draws extensively, is right that the woman is a "Hellenized Phoenician," then she is herself the result of a complex colonial history. Tyre's origins were of course Phoenician—Northwest Semitic—but after it was conquered by Alexander in 332 BCE, it became part of the Greek cultural world of the eastern Mediterranean. Much of the original population was sold into slavery, and the city was resettled, ultimately achieving a distinguished status as a center of Greek education and learning. By the time of Mark's story imperial power had moved on, and Tyre was part of the Roman province of Syria and now at the eastern periphery on a new imperial power. Phoenician may well have already been dying out as a spoken language by the time of Jesus, and much of our access to Phoenician history and tradition comes through the highly Hellenized work of Philo of Byblos. Nevertheless, Tyre retained an unusual degree of independence within the empire, and it would be a mistake to think that older elements of Phoenician culture were wholly eclipsed by a new Greek identity (Millar 1983 and especially Nitschke 2011). All in all, it seems likely that a woman of the sort Mark describes would have been yet more multidimensional than Kinukawa suggests.

---

people who first discovered the British Islands and made them known to mankind at large, the people who circumnavigated Africa, and caused the gold of Ophir to flow into the coffers of Solomon."

THE WRONG KIND OF ISLAND? 189

The editors in their introduction note that the volume does not make much of either "bad" islands (like Robben Island) or imperial islands like Britain or Japan. This essay is the exception, since its introduction and conclusion bring Kinukawa's experience as an imperial islander to the fore. Unlike most of the authors in the collection, Kinukawa takes no pleasure in her island heritage; the dominant note seems to be one of shame or regret. But this regret is productive in that her awareness of the dynamics of Japan's relationship with its colonial possessions provides the initial key to unlock the meaning of Mark's story and find in it a positive message for both sides. There is a sharp reminder here that if islanders are readers, it would be a mistake to assume that all islanders will read in similar ways.

Kinukawa has perhaps one advantage over imperial islanders from the British Isles, and that is that the Bible is not itself a fundamental resource for Japanese ideology as it has been in Europe. One of the most consistent targets of this volume is the idea that Western conceptions of islands and islanders have failed to do justice to the reality of island life and experience. The Bible itself is not innocent in this process and if, like Kinukawa, I want to reflect on my place as an imperial islander, then I may also need to look rather more suspiciously at the role biblical texts have played in the Western imagination of islands. As a counterweight to Kinukawa's experience, one such example from the age of exploration comes from the story of the first Japanese embassy to Europe. The historian J. F. Moran (1993) describes how between 1582 and 1590 four young Japanese converts were brought by the Jesuits on a tour of Europe where they travelled through Portugal, Spain, and Italy, meeting two popes and many other important religious and secular leaders, most notably Philip II of Spain, whose empire at that point stretched from the Americas to the Philippines. In September 1585 the Japanese ambassadors were greeted at Zaragoza Cathedral by the singing of *Reges Tharsis et insulae munera offerent*, the Epiphany Offertory drawn from Ps 72:10–11:

> May the kings of Tarshish and of the isles
> render him tribute,
> may the kings of Sheba and Seba
> bring gifts.
> May all kings fall down before him,
> all nations give him service.

Since this was in September rather than January, Epiphany itself was several months away. The event had a clear theological and political symbolism, underscoring Philip's claim to be a Christ-like world ruler and emphasizing the subordinate and tributary status of these islanders from afar. Indeed, the imperial aim is made quite explicit in a general statement of the reasoning behind the embassy made by its organizer Alessandro Valignano:

> to make the Japanese aware of the glory and greatness of Christianity, and of the majesty of the princes and lords who profess it, and of the greatness and wealth of our kingdoms and cities, and of the honour in which our religion is held and the power it possesses in them. (Moran 1993, 8)[3]

As with Alexander's conquest of Tyre, this case serves as a reminder that even imperial islands can also end up as the objects of other people's imperial ambitions! It is striking that Tyre's loss of independence was also marked by Alexander's building of a causeway from the mainland to the city and that with the loss of its independence came the loss of its island status.

In a 1993 newspaper article, the Scottish columnist John Macleod reflected on the danger that such causeways and bridges pose to the distinctive island identity of the Scottish Western Isles. Writing of proposed bridges from the tiny island of Scalpay to the larger island of Lewis and Harris and from the Isle of Skye to the Scottish mainland, he claimed that many Hebrideans "are to lose even their island identities, a whole mindset changed by the inexorable sweep of the bridge" (Macleod 1993). It is these Hebridean identities that occupy Grant Macaskill, as he introduces his readers to the practice of Gaelic psalm singing in the Western Isles of Scotland. As well as being the only part of Scotland that retains a substantial number of Gaelic speakers, these islands are home to Scotland's most traditional reformed churches, in which the rule has long been for the only music in worship to be unaccompanied psalm singing. Macaskill describes the unique musical form of Gaelic psalmody and goes on to link the theology of place that is ubiquitous in the Psalter with the sense of place that belongs to island communities. He writes of a "geographical translation" of psalmic places in which the biblical wildernesses and hills are connected in the singing with their equivalents on Lewis or Skye. "The act of singing," he says, "becomes an exercise of the imagination, mapping psalmic

---

3. I have undertaken a broader study of the use of Ps 72 in a range of imperial contexts in Mein 2009.

language onto island space, as an articulation of the placed encounter with God" (Macaskill in this volume, 104).

These connections with specific known places not only emphasize God's engagement with the natural world, but also keep the specific places of the island experience constantly within the islander's horizon. I was especially struck by his description of the monument to the Iolaire tragedy, which stuck Lewis at the end of the First World War. On New Year's Day 1919 the ship Iolaire, full of men returning to the islands from their wartime service, sank only minutes away from the safety of Stornoway harbor, with the loss of nearly two hundred island men. The Iolaire memorial describes the event in English and then offers a Gaelic quotation of Ps 77:19: "Thy way is in the sea, and in the waters great thy path." Macaskill argues that this is understood to affirm God's oversight not only of the sea in general, but of this particular known stretch of it. With this and other examples, he makes an effective case that the psalms have contributed not only to a theology of place, but also to a strong and distinctive island identity.

The psalms do seem to have played a particularly powerful role in the religious identity of Scottish Gaelic speaking communities. It is not only the fact that the Western Isles are islands that have preserved this distinctive Gaelic-speaking Presbyterian culture, but also that a specifically biblical idiom has been able to thrive in Gaelic in this particular environment and so inform an island identity. There is surely more to be said here about both the vulnerability and vitality of an identity based on language and the role of biblical translation in shaping and preserving communal identities.[4] One might argue that, by comparison with some of the other linguistic groups within the British Isles, such as speakers of Cornish or Scots, the Gaels had it pretty good.[5] There was an early Gaelic translation of the Church of Scotland's prayer book, the Book of Common Order, and the first fifty psalms were available in 1659, with the rest of the Psalter following behind in 1694. The translation of the rest of the Bible was more hit and miss: an Irish version existed from not long after the Reformation, and this was replaced by a Scots Gaelic one around the beginning of

---

4. See the essay by Ma'ilo in this volume.
5. For reference, Scots Gaelic is part of the Celtic group of languages (along with Irish, Manx, Welsh, Cornish, and Breton), whereas Scots (sometimes Lowland Scots) is a Germanic language that developed, like English, from Anglo-Saxon, but with its own distinct grammar, vocabulary, and distinguished literary tradition.

the nineteenth century. But literacy levels remained low in this predominantly oral culture, and evidence for the speedy and widespread adoption of any of these translations is sparse (Meek 2002). It seems to have been especially the Psalter, remembered and sung every week, that put biblical language at the heart of Gaelic culture.

And here my reaction is again shaped by my own island identity. For me, Macaskill's subject matter is much closer to home, but in some ways no less alien than the other two. I have already identified myself as a British islander, but lest we assume that one island equals one identity, I need also to identify myself as a lowland Scot, brought up in an English-speaking metropolitan center with very little contact with Gaelic culture. And it is striking to me that vitality of the Gaelic biblical tradition that Macaskill describes contrasts sharply with the situation in the Scottish Lowlands, where there is still no complete published translation of the Bible into the Scots language, and hence no distinctive biblical idiom with which to construct a Scots theology of place. The first translations of the psalms into Scots were not until the second half of the nineteenth century, by which time Scots was almost dead as a literary language and awaiting revival. The first vernacular Bibles printed in Scotland were English ones, with a version of the Geneva Bible in 1579 followed by another edition in 1613, and in due course the KJV in 1640 (Gribben 2009, 11–12).[6] Unlike Gaelic, Scots stands linguistically very close to English, and both the Geneva Bible and the KJV would have been more or less comprehensible to most Scots, especially if accompanied by a Scots sermon. This obviated the need for a translation into the language that people actually spoke and contributed heavily to the demise of Scots as a language of education and literature, not least since from 1579 every substantial Scottish householder was required by law to own a copy of the Bible in *vulgare langage*, which by default had to be English.[7]

---

6. It is of course a deep irony that this monarch responsible for introducing the definitive English version of the Bible was himself a Scot, who wrote extensively in the Scots language as well as in English.

7. The situation is exemplified beautifully in Robert Burns's poem, "The Cotter's Saturday Night," which describes a typical Saturday evening in the house of a Scottish tenant farmer. It starts off in Scots and reaches the point where the patriarch reads from the "the big ha-bible, ance his father's pride," at which point it "slips into formal English in the reading, the psalm singing, and the prayers" (Murison 1989, 11). It reads, "And 'Let us worship God!' he says with solemn air." David Murison (1989, 12) also goes on to note the similar move in another of Burns's best-loved poems

To return to the Western Isles, in so far as Macaskill identifies his own reading position at all, it is as an islander rather than a Briton, and it is noteworthy that Macaskill never mentions Britain as part of the identity of the Western Islander identity. The contrasts are always between island and mainland, or island and the rest of Scotland. And in this he participates in the general trend of the volume to place islands and islanders at the periphery, the margins, and to see them as minoritized and anti-imperial.

But, like Kinukawa's Syro-Phoenician woman, Macaskill's islanders do belong to a rather more complex web of both colonized and colonizing identities. The Iolaire disaster is instructive in this case, since it is important to remember that the ship was bringing back men who had been fighting on for Britain and its empire. Highlanders and islanders had provided much of the backbone of Britain's fighting strength throughout the past centuries of imperial expansion. Indeed the story goes that it was a Lewisman named Robertson who raised Nelson's famous signal at Trafalgar: "England expects every man to do his duty" (see Paxman 2007, 44), and kilted highlanders ("petticoated devils") provide perhaps the most typical images of the nineteenth-century British soldier (McNeil 2007). It is therefore no surprise that when war broke out in 1914, a far higher proportion of the male population of Lewis joined up than the national average.[8]

Here I should like to come back to the problematic role the Bible has played in constructing a range of island identities and to a couple of examples from the psalm paraphrases of Isaac Watts (1674–1748), which represent two different strands of British imperial thinking about islands. As I mentioned before, Macaskill describes a process of "geographical translation" whereby psalm texts are reinterpreted in the light of the island environment of the Western Isles. Something rather similar is going on in Watts's paraphrases, but one that is far less easy to see as innocent or wholesome.

Watts's great missionary hymn "Jesus Shall Reign," composed in 1719, is a version of Ps 72, whose most obvious interpretative features are a strongly christological understanding of the royal figure at its center and

---

"Holy Willie's Prayer," a biting satire on hypocritical Calvinist piety. Willie's public and respectable prayer is in liturgical English, "the language of the Bible, the church, the state and all its works," while his more earthy desires are expressed in Scots.

8. For Lewis itself, nearly 7,000 men out of a total population of around 30,000 served in the forces. The 1,150 war dead also represented double the national average (MacLeod 2010).

an equation of the extent of his kingdom with the growth of Christianity throughout the world. Over the years the hymn became vastly popular, and Michael Hawn (2005) goes so far as to claim that it became an epitome of the expansionist theology of the missionary movement. The islands make their appearance in the original second verse of the hymn, which along with the third, is rarely sung today. They paraphrase Ps 72:10–11 as follows:

> Behold the islands with their kings,
> And Europe her best tribute brings;
> From north to south the princes meet,
> To pay their homage at His feet.
> There Persia, glorious to behold,
> There India shines in eastern gold;
> And barb'rous nations at His word
> Submit, and bow, and own their Lord.

Here, as in the example of the Japanese ambassadors, "the islands" fulfill the stereotype that this volume is so at pains to undermine. Islands are remote, tameable, and tribute bearing. Admittedly, islands are not uniquely exploitable, since Watts has replaced Tarshish, Sheba, and Seba with Europe, Persia, and India. The latter two, although not islands, were certainly places where Britain in 1719 was expanding its political and economic interests rather faster than its missionary activities. Hawn asks whether it is too much to think that Watts's singers were singing subconsciously.

> [Britain] shall reign where'er the sun
> Does [her] successive journeys run;
> [Her] kingdom stretch from shore to shore,
> 3Till moons shall wax and wane no more.

The second kind of geographical translation that Watts indulged in was quite explicitly to replace "Israel" with "Britain" in a number of psalms. These psalm paraphrases, as John Hull has pointed out, express a nationalist theology that equates Britain with Israel and sees Britain's growing influence in the world as the fulfilment of Christ's universal reign (Hull 2002). It is perhaps clearest in the paraphrase of Ps 47:

> In Israel stood his ancient throne,
> He loved that chosen race;

But now he calls the world his own,
And heathens taste his grace.

The British Islands are the Lord's,
There Abraham's God is known;
While powers and princes, shields and swords,
Submit before his throne.

The God of Israel has transferred his allegiance to Britain, and the conversion of "heathens" is paralleled by the subjection of "powers and princes." Israel is transformed into an archipelago, and the "British Islands" stand at the center of both the divine and the human economies, with the baleful consequences that provide much of the context for the collection of essays in this volume.

The wealth of island experiences that lie behind the essays in this collection have set me to reflect for the first time on my own island identity, and as a British islander, I cannot but share some of Kinukawa's deep anxiety about my own islands' imperial and colonizing past. And that anxiety is compounded by examples like that of Watts above, which show that biblical interpretation itself had a role to play in the construction of an imperial island ideology. I nevertheless hope that there is value in taking time to name some of the problematic aspects of the relationship between island and Bible, and that even those of us from the wrong kinds of island might thus contribute to the development of island hermeneutics.

Works Cited

Champion, Timothy. 2001. "The Appropriation of the Phoenicians in British Imperial Ideology." *Nations and Nationalism* 7:451–65.

Gribben, Crawford. 2009. "Introduction." Pages 1–18 in *Literature and the Scottish Reformation*. Edited by Crawford Gribben and David George Mullan. St. Andrews Studies in Reformation History. Farnham, UK: Ashgate.

Hawn, C. Michael. 2005. "Singing with the Faithful of Every Time and Place: Thoughts on Liturgical Inculturation and Cross-Cultural Liturgy." *Colloquium Journal (Yale Institute of Sacred Music)* 2. http://ism.yale.edu/sites/default/files/files/Singing with the Faithful of Every Time and Place.pdf

Hull, John M. 2002. "From Experiential Educator to Nationalist Theologian: The Hymns of Isaac Watts." *Panorama: International Journal of Comparative Religious Education and Values* 14:91–106.

MacLeod, John. 1993. "Taking on a Peninsular Outlook." *Herald*, August 7. http://www.heraldscotland.com/sport/spl/aberdeen/taking-on-a-peninsular-outlook-1.747638.

———. 2010. *When I Heard the Bell: The Loss of the Iolaire*. Edinburgh: Birlinn.

McNeil, Kenneth. 2007. *Scotland, Britain, Empire: Writing the Highlands, 1760–1860*. Columbus: Ohio State University Press.

Meek, Donald. 2002. "The Pulpit and the Pen: Clergy, Orality and Print in the Scottish Gaelic World." Pages 84–118 in *The Spoken Word: Oral Culture in Britain, 1500–1850*. Edited by Adam Fox and Daniel R. Woolf. Manchester: Manchester University Press.

Mein, Andrew. 2009. "Justice and Dominion: The Imperial Legacy of Psalm 72." *Bangalore Theological Forum* 46:143–66.

Millar, Fergus. 1993. *The Roman Near East, 31 BC–AD 337*. Cambridge: Harvard University Press.

Miscall, Peter D. 1992. "Isaiah: New Heavens, New Earth, New Book." Pages 41–56 in *Reading Between Texts: Intertextuality and the Hebrew Bible*. Edited by Danna Nolan Fewell. Literary Currents in Biblical Interpretation. Louisville: Westminster John Knox.

Moran, J. F. 1993. *The Japanese and the Jesuits: Alessandro Valignano in Sixteenth Century Japan*. London: Routledge.

Murison, David. 1989. "The Two Languages of Burns." Pages 1–14 in *In Other Words: Transcultural Studies in Philology, Translation, and Lexicology Presented to Hans Heinrich Meier on the Occasion of His Sixty-fifth Birthday*. Edited by J. Lachlan Mackenzie and Richard Todd. Dordrecht: Foris.

Nitschke, Jessica. 2011. "'Hybrid' Art, Hellenism and the Study of Acculturation in the Hellenistic East: The Case of Umm el-'Amed in Phoenicia." Pages 87–104 in *From Pella to Gandhara: Hybridisation and Identity in the Architecture of the Hellenistic East*. Edited by Anna Kouremenos, Sujatha Chandrasekar, and Roberto Rossi. BARIS 2221. Oxford: Archaeopress.

Paxman, Jeremy. 1999. *The English: A Portrait of a People*. London: Penguin.

Scott, Jonathan. 2011. *When the Waves Ruled Britannia: Geography and Political Identities, 1500–1800*. Cambridge: Cambridge University Press.

Theissen, Gerd. 1991. *The Gospels in Context: Social and Political History in the Synoptic Tradition*. Minneapolis: Fortress.

THIRD WAVES

Third Wave Reading

Elaine M. Wainwright

Some would claim that a third wave is generally bigger than the two that go before it, but that is not so here. There has been a huge wave that has brought with it all manner of newness that it leaves on the beach: different colored and differently shaped shells and stones, beautifully colored seaweed, objects that are strange and new whose earlier destination is unknown, and even a piece of exquisite driftwood. Over this washes a gentle and subtle second and third wave. These two are much smaller than the first wave, and they do not disrupt or disturb what the first wave has deposited. One can walk among the deposits as they are gently touched by the second and third wave. But one must take care with such walking—to tread heavily, one could destroy an exquisite piece; to be inattentive, one could miss the most beautiful configuration; and to engage another deposit, one may need to bend and take it up, turning it around and viewing it from different angles.

To take up and engage with this extraordinary collection, *Islands, Islanders, and Bible: RumInations*, evoked for me the image(s) above. They are born of my islander identity, albeit from the large island of Australia (Boer in this volume notwithstanding). They are also born of a love for the borderland place between sea and land, the island's beaches, and that experience of walking a strewn beach with the expectation of finding the beautiful and the new. I am very grateful, therefore, for the invitation to participate in the third wave, responding to four selected articles in the collection: those of Margaret Aymer, Mosese Ma'ilo, and Hisako Kinukawa together with the opening chapter, "RumInations."

It is in "RumInations" that the reader begins to encounter the wonderful array of new insights thrown up upon the beach of contemporary biblical interpretation. The play upon the word *rumination* in the different sections images well the regurgitation that rumination evokes, the coming back to, the reengagement with. The reader is also invited into the calm and lengthy consideration that is rumination. The ruminants, both writers and readers, are islanders and biblical scholars—an exciting new configuration in the academy (but not in the islands).

As the editors seek to mark out a space that is rumi-/roomy enough to allow for an island perspective, they engage critically with Western culture's construction of the island and islanders. Many contemporary biblical scholars (postcolonial, feminist, minority, and queer to name but some) are familiar with the significance of critical engagement with the cultural and academic perspectives that dominate in the discipline. This study draws the constant attention of readers to a new space from which biblical interpretation is emerging—that of the island. This is the space that shapes islanders but awareness of the significance of such space and the experience it generates has also given rise to a new academic discipline or thread, "island studies."

Movement between the known and the new continues as periphery and margin, categories particularly familiar to a range of contextual biblical scholars are refracted very differently through the islander lens—"the notion of peripherality and its correlate of alter/native presents the opportunity of seeing island spaces not simply as responses or write-backs to nonislands spaces, but rather spaces of originality and innovation" (Davidson, Aymer, and Havea in this volume, 7). The challenge for the future of such explorations will be that of "originality and innovation" rather than "write-backs."

As the island location is further explored in rum-I-nations, the second major section, attention turns to new biblical interpretations that are possible from island locations. There is space, however, for the Bible to speak back to islands and islanders in a very creative dialogue—this in its turn would throw up more treasures from the sea in an extended first wave or a new first wave beyond this volume itself. A third section attends to the nations and raises the question as to what island and nation have to do with the "island hermeneutic thing" (13–19)—an area of possible contestation into the future of biblical studies. A creative and playful close—biblical texts are like islands, and readers are like islanders—seeks to address this, throwing up yet another piece that is washed by this third wave revealing the newness that island and islander ruminations will offer the future of biblical scholarship.

In turning to the three interpretive essays, I will engage with them through the lens of three of the images in my opening paragraph. The first

I articulate this way: to tread heavily could destroy an exquisite piece—and it is through this image that I read Aymer's rumination, "Islandedness, Paul, and John of Patmos." She herself treads lightly around the exquisite lens she has created, namely "islandedness." There is, however, also a strength, a power in this concept as she claims that islandedness can function as a "wedge" that might "split open" in new ways the study of the biblical text and its ancient world (in this volume, 25).

As Aymer establishes her reading site, her reading lens, and herself as reader, she also moves exquisitely around and within her own complex history/experience and relationship(s) with islandedness. It is much more than location and even more than history. Indeed she does a new thing—she uses islandedness as a tool "to think with" (26). In this way, she explores and explodes any notion of the simplicity and/or the colonial glorification of island living. Rather the complexity of movement from, to, and within the islands makes for a range of identities that coexist in complex ways. In this she provides a challenge not only to island readers but to all contextual biblical readers and interpreters. Location is a complex reality that requires careful analysis for its nuanced perspectives, and so Aymer challenges islander (and other contextual) readers to the careful analysis of the counterpointed positioning that is location. Islandedness is the exquisite piece that has been thrown up by the first wave, and the third washes over it in a way that draws attention to its beauty.

After her reading of Paul and John of Patmos through an island lens (and later her call to extend such readings), Aymer returns to some of the tantalizing questions that were becoming visible in the opening sections of her article. These are hermeneutical and interrogate all readers. Having established islandedness as a significant hermeneutical lens, Aymer allows this claim to return to interrogate herself—is there such a reading site, such a perspective that allows for new interpretations to emerge? In this she raises a challenging question for all contextual interpretations: what are the parameters that make for unique interpretations? Perhaps her most significant challenge is that of the range of other categories within a hermeneutical lens—that of "gender, sexual orientation, class, age, ability, migrant status, et cetera" (35). It is only an exquisitely sensitive movement through these waters, like that of the one who negotiates the third wave as it washes over the exquisite piece on the beach, that will enable an islander hermeneutic and islandedness to emerge in all its complexity and to provide a new reading lens for biblical interpretation.

In reading the Ma'ilo's interpretative article, "Celebrating Hybridity in Island Bibles: Jesus, the *Tamaalepō* (Child of the Dark) in Mataio 1:18–26," I propose to do so through the lens of the second aspect of my opening imagery: to be inattentive one could miss the most beautiful configuration. Mosese Ma'ilo invites readers into the complex and indeed hybrid world of translation as vehicle of colonization but also a vehicle of the indigenization of Christianity. The lens of hybridity that Ma'ilo employs is an ideal channel for the attentiveness that my metaphor calls for so that one does not miss a most beautiful configuration. Indeed, in his closing sentence of his analysis of the hybrid nature of the translated text, Ma'ilo states that "island Bibles represent a language *in-between* ... a space where foreign and local symbols are brought into the harmonious creation of new concepts and meanings" (67)—a new, most beautiful configuration.

Hybridity as a hermeneutical lens weaves its way through Ma'ilo's reading of Matt 1:18–25 as well as his closing reflections. Like Aymer's islandedness, Ma'ilo's use of "hybridity" provides a very nuanced place to stand, a place which gives rise to a most beautiful configuration. He takes his readers on a two-way journey. He demonstrates how translations speak, presenting the pregnancy of Mary in an almost simplistic way. But then as reader, he speaks back to the text demonstrating the shift in Jesus's identity pre- and postresurrection using Samoan concepts. In this way he has, perhaps, answered his own question: "Where do we locate culture or native traditions in the hermeneutical activity?" (74). It is through such reflection on one's hermeneutical activity together with interpretations infused with and informed by island cultural concepts that new and emancipatory interpretations, new and colorful configurations, will be washed up on the beach of contemporary biblical interpretation.

In order to engage Kinukawa's reading of the island of Tyre through the lens of the "exploitation of peasants in the regions of Tyre and Galilee" (135–45), one may need to bend and take it up, turning it around and viewing it from different angles. First, she stands very tentatively in her space, the island of Japan, which she characterizes as "gobbling up products and people in Asian countries" (136). This is a very different starting point from that of Aymer and Ma'ilo; it is that of the colonizer rather than the colonized, but a colonizer who is attentive to power, economic, and cultural differentials and where she and her island(s) stand. From her island location, Kinukawa turns her reading lens on islandedness in the biblical text, as Aymer suggests is one possible approach for island readings. The island she reads is that of Tyre as it is narrated in Mark 7:24–30.

Kinukawa takes up her new recognition that Tyre was an island, and she turns it around and views it from different angles—historical, geographical, and economic to name but three. Its affluence, in contrast to the poverty of its Galilean hinterland, Kinukawa reads as the block which prevents Jesus from seeing the Syro-Phoenician woman as other than Tyrian. It is only when Jesus recognizes the differences that exist within the social fabric of Tyre itself that he can see her and her daughter as ones in need. An island reading enables a reader to enter an old text from a new direction, to ruminate on that text, and to call forth new meanings.

This third wave reading has been both privilege and pleasure like that of walking attentively along an island edge that has received the riches the ocean gives up. The experience has been transformative as I have encountered challenging and evocative ways of engaging islandedness, my own and that of others from a wide range of island locations and experiences. Exploration of the "island hermeneutic thing" has been both subtle and explicit. It has not isolated interpreters one from the other but rather engaged readers in its complexity. Ruminating on that has only just begun, and biblical scholars can look forward to many more exquisite finds along the beach after a range of new first waves to come. Accompanying new hermeneutic finds will be interpretations as islanders engage their Scriptures in new and enriching ways. Indeed, I look forward to many second and third wave experiences into the future as I explore the riches that islander interpretations will deposit on its beaches.

Thinking on Islands

Daniel Smith-Christopher

Hearing the Song

In Aotearoa/New Zealand, the old tradition for meeting the Maori is that you approach a Marae (a Maori communal sacred space and settlement) with caution to indicate your peaceful intentions. You wait at a distance but within sight. In previous ages, the warriors would caution you. If you responded in peace, a welcome would be sung by the women. These days, the warrior's warnings are largely for the dinner shows (and the Rugby world champion, "All Blacks"). Modern Maori warriors are priests, businesspeople, and lawyers, male and female—and the fights are in Parliament ... but the singing welcome is a continued legacy.

As a visitor to the Islands, Islanders, and Bible group at the Society of Biblical Literature, I have learned to wait to be invited. When I hear "Haere mai, haere mai," I know I may proceed. It is important for me to clarify, therefore, that while my attendance, support, and interest in the group represented in this book was at my own initiative, my actual involvement in this book has been by invitation. Having heard the singing of "Haere mai," I want to respond by expressing my appreciation to the steering committee, and particularly Jione Havea, for this invitation, as well as my appreciation to the four colleagues whose fascinating essays I am invited to engage in dialogue. It is my conviction that this work is the beginning of something very important.

"Third Wave" Responses

In their opening essay, Steed Vernyl Davidson, Margaret Aymer, and Jione Havea survey a variety of fascinating issues raised by the idea of islanders reading the Bible and at the same time demonstrate the importance of a volume of contributors gathered around these issues. Along the way, they ask a number of important questions: "Does 'place' matter for interpretation, and if it does, how does it matter?" (8); "to what extent do 'mainlanded' theories threaten to drown out other sorts of questions that derive from the peculiarities of insular life?" (10); and "how can questions about the nature of insularity be used to think about biblical writings, communities, and formations?" (10–11). Toward the end of the essay, the editors state that "island hermeneutics joins arms with other hermeneutics of suspicion and of resistance in advocating minority (islandish) and minoritized subjects and interests" (18).

If it is true that the location of the interpreter of a text demonstrably influences his or her readings of any text, including the Bible, this is only the first step toward the conclusion that a volume such as this is of critical importance. The first question, regarding the importance of place for interpretation, could be answered, after all, with a negative conclusion—for example, "Yes, reading from different places does matter, and this kind of bias needs to be scrupulously avoided for a more scholarly approach." I do not agree with this view, but I have read it, and heard it, innumerable times. Such a possible negative reply effectively brings on the second and third questions about ideas that may uniquely derive from islander perspectives—ideas that may perhaps offer important new insights about the texts we are all involved in working on.

For those of us convinced that some of the most important insights about historic texts come from an ability to ask new questions, then it seems obvious that colleagues in biblical studies who can draw on underrepresented perspectives, cultures, and points of view can therefore also offer something of tremendous value—*new questions that the rest of us simply would not think to ask!* The possibility that islanders may lend voices to what the editors call "minority and minoritized subjects and interests" is already worthy of support (18). But there is more. Davidson, Aymer, and Havea offer more than simply a few more underrepresented voices (such a list would be depressingly long ... and presents a serious challenge to those of us engaged in higher education). Davidson and his colleagues show us that this particularly articulated voice—those whose life experience includes a self-identity characterized by significant life experience on an island—will offer equally and particularly articulated new questions.

One of the benefits of reflecting on the editors' programmatic introduction, then, is how it launches all of us on a new journey. Since my own very brief island experience is focused mostly on my continued involvement with, and fascination with, New Zealand/Aotearoa, the launching imagery that comes to mind are the large "Waka," the large Maori canoes capable of holding dozens and dozens (up to 130 feet long!). In other words, there is room for lots of us to go along! There are so many islands and island cultures, not to mention the invited guests.

It is clear, however, that the introductory essay identifies potential problems as well. There are certainly some common experiences that come from the geographical similarities of isolated existence of islands, but even here, one must differentiate between those islands that are so connected as to be virtually an undifferentiated part of mainlands (Manhattan or *modern* Venice or Catalina) and those whose experience of extreme isolation is so much a part of the self-identity of its residents (the Pacific islands, for example) that the sea has been incorporated into their identity to such an extent that they simply do not think in the same terms of isolation as mainlanders do! After all—islanders know the seas as mainlanders know the roads. It seems to me, however, that experiences on this level can be generalized so much that the value for interpretation of texts is almost entirely metaphorical (How are texts also isolated? In what ways do biblical texts reflect times of relative isolation from an awareness of larger kingdoms, through times when those kingdoms swallow up elements of the homeland after 587 BCE? or 70 CE? etc.). These are interesting questions, of course, which an islander perspective already raises about historical-critical questions.

However, I would suggest that the editors' arguments about islands as almost inevitably on the margins of "mainland" thinking, nevertheless, hold a key to an even more important insight—the unique perspective of these *different* island peoples. Here, each of the contributions begin to be specific, and in their specificity begin to offer even more valuable gifts of insight and the promise of much more. For me, my involvement as a supporter of the Islands, Islanders, and the Bible section at the Society of Biblical Literature has its' roots in my own obsessions with the utterly fascinating history of New Zealand.

New Zealand is a group of islands, but unique to *these* islands is the sheer genius in the Maori appropriation of Christianity and the Bible: (1) Te Whiti's *pre-Gandhi* experiments with faith-based nonviolent resistance to colonial abuses; (2) Te Kooti's reflections on Scripture and revolution; (3) Wiramu Tamihana's ruminations on biblical royalty, which led him to propose and support an idea entirely without Maori cultural precedent—a unifying "Maori King" movement, which he believed would lead to peaceful Maori unity and empowered coexistence with Europeans. Much has been written (and still needs to be written) about the brilliance of the African American appropriation and reinterpretation of Christianity and Scripture, but the Maori experience is only beginning to be understood and appreciated. Bronwyn Elsmore admirably started the task with a general introduction to the topic (*Like Them That Dream: The Maori and the Old Testament*, 1985), and Judith Binney's far more analytical and focused analysis effectively points to Te Kooti's work with Scripture in her magisterial historical and biographical work (1997: 289, 318, 426, 528–34). Neither is a Scripture scholar, however, and so the *details* of Te Kooti's exegesis, and other Maori leaders of the nineteenth and twentieth centuries, remain as yet unpolished, but still beautiful, pieces of greenstone.

It is also on the "land of the long white cloud" (the literal meaning of "Aotearoa") that we find equally engaging unique contributions of Pakeha theologians and activists ("Pakeha" is the term used for "European New Zealanders") whose engagement with the land of New Zealand has given rise to some startlingly provocative ideas. Here we find the missionary turned pro-Maori activist Octavius Hadfield, who was hated by rapacious settlers; Rev. Ormand Burton, the WWI hero who turned into an antiwar pacifist preacher; the poetry of James Baxter who tried to start a Christian commune on Maori land where he was invited to live; and preeminently the stunning painting of Colin McCahon, who introduced biblical texts and debates directly into his highly controversial (to this day) Modernist paintings. Contemporary work continues apace, especially among biblical studies faculty at the Universities in Auckland, Otago, and Wellington. I have had enough conversations with my friend, Dr. Steven Jennings of Kingston (see, now, Jennings, 2007) to know that an equally impressive list would take him only a few moments to assemble for Jamaican history and experience.

My point in this brief (and, quite admittedly, selective) survey of issues, is to emphasize that *each* of the island experiences will have their unique gifts. Davidson, Aymer, and Havea's essay helps us to realize the importance of supporting any attempt to unite and support the work of islander biblical scholars, who offer not only generalized insights from a somewhat formulated common island experience, but especially the unique treasures of different island cultures and histories (with an ironic nod to Robert Louis Stevenson, who did not realize that the real treasure is not buried *in* the islands, it lives *on* the islands).

Althea Spencer Miller's reflection on "creolization" as a social and political response to conquest is also very suggestive. At first, creolization is presented as the strategy of survival and continuance of a people buffeted by forced transportation and nearly constant cultural interference. Like an island resiliently standing up to tropical storms pounding away, creolization is presented as a key to persistence and resistance (83).

However, it strikes me that creolization is also an incredibly generous response by a people with very few historical reasons to be generous, specifically, to Europeans. It is generous in that it is inclusive. If creolization is a creative mixing—creative in that islanders (in this case Caribbean islanders) create this form of communication—then it is the mixing that is generous—our words are included. In fact, Miller's methodology is typified throughout by facing nonislanders with a rather undeserved generosity. Furthermore, Miller states that the essay's conclusions will be open because of a refusal to draw firm conclusions and "expresses a commitment to a communal rather than an idiosyncratic and individualistic claim to sagacity, regional or otherwise" (78). Even the topics are open for discussion, not presented as a *fait accompli* with the implied challenge that "it is whole and complete ... just try to challenge me."

Miller's reflections on creolization raised questions for creative possibilities in our interpretation of biblical texts. While there are certainly moments of xenophobic rejection of all things "other" in the Hebrew traditions (one thinks of Levitical purity as it translates into Ezra's rejections of anyone other than "the Holy Seed," so Ezra 9:2; also the worries about Hebrew children speaking the wrong language in, for example, Neh 13:24), there are interesting other possibilities when noting that the book of Proverbs includes aspects of Egyptian wisdom, which may represent a kind of *philosophical creolization* that signals an openness to the value of other's thoughts and traditions (cf. Paul's Athenian sermon in Acts 17). Under Miller's suggestions, it seems to me that a reexamination of many cases in the Hebrew Scriptures where Hebrews converse with non-Hebrews needs to take place with the new goal of seeking examples of any kind of "creolization"—any openness in those conversations that suggest something of the creative generosity of Caribbean engagement with others. So, "conceptual" creolization holds a great deal of promise, similar to discussions of "hybridity" in postmodern idiom. But even in a more technical linguistic sense, Miller's suggestions raise equally provocative possibilities. In fact, many discussions of particular linguistic forms of Hebrew are conducted in the somewhat sterile terms of influence from this or that Semitic variation, without much thought to the *social* processes that may be behind grammatical or patterned linguistic changes. Finally, are Jesus's occasionally noted phrases of Aramaic in the Gospels an indication of the language of his actual daily usage, or are they rather indications of his *occasional* use of phrases, or grammatical mixing, a kind of first century "creole" that comes from occupying a particular island of Roman Imperial space? How *did* he address the Roman Centurion, or Pilate, for that matter? It would seem to me that Richard Horsley's (1997, 2002) reflections on the early Christian appropriation of Roman phrases or terms (even "good news") could be examined from the perspective of a social and theological creolization.

Finally, I am struck by Miller's suggestions that the importance of oral communication cannot be marginalized by an exclusive dependence on written communication. Would island biblical studies benefit from exploring different digital forms for academic publication that might include video conversations that are treated with the same reverence as peer-reviewed writings? Should the peer review of such a conversation be *further conversation*? This context creates new meaning to the common typographical error: in the island context, a "peer" review can be a "pier review"—conversations on the shore! If conversation, songs, dances, and even recipes (Miller's essay in this volume, which includes meal descriptions, makes us hungry as well as informed!) are integral to island identities, then appropriate "journals" must be rethought.

There is a striking asymmetry in comparing Miller's essay to Grant Macaskill's essay. Miller is reflecting on the experience of an island that was constantly visited (by guests both welcome and unwelcome) as part of the "New World" and therefore becomes a place of creative mixing. It is the interaction that gave rise to what Miller's islands offer the student in her careful presentation. Macaskill's reflections, on the contrary, come from an island experience quite the opposite—the very isolation from constant visitation is what gives rise to the treasures it offers the serious student and even the (much more occasional) visitor.

I was also present when Macaskill presented his amazing essay on Gaelic psalmody in the Western Isles of Scotland. I may have been the only one in the room, other than Macaskill himself, who has actually stood in the presence of the "Callanish Stones" of the Isle of Lewis, but I will never forget that day (years ago now) in the driving rain, determined to see them. Unlike the more famous Stonehenge (but not less impressive), one could actually approach the Callanish Stones—one of the benefits of their isolation from the crowds. On the day, I was very much alone!

To read Macaskill speak of the determination of the islanders maintaining their devotion to the singing of psalms reminded me of seeing those stones in the driving rain—geological metaphors of the very people Macaskill was discussing. Macaskill's reflections on how psalms are rendered by these island people as referring not only to the biblical lands but also their own—and the interpretive traditions that inform these exegetical experiments—reminds us of the urgency of continued studies like Macaskill's important work. Here we must "gird our loins" against those who would dismiss such efforts as hardly worth the academic time and grant-seeking to support such work. Macaskill's essay singularly puts an end to such nonsense. But another example can further illustrate the importance of work on the isolated *as well as* the populous islands.

The Sea Islands off the southeastern coastlines of the United States presents a similar case of how isolation served to preserve cultural (and even theological) traditions. Although obviously taken quite seriously by African Americans themselves from the time of the Wesleyan revivals among slaves and growing conversion to Christianity, the African American music known universally as "the Spirituals" began to be written about and discussed by European and European-American writers as early as 1867. In that important year, the first published *collection* of spirituals (i.e., *Slave Songs of the United States*) was edited by William Francis Allen, Charles Ware, and Lucy McKim (who married the second son of William Lloyd Garrison), largely based on field collecting in the Atlantic coastal African American communities and especially the Sea Islands (also known as the "Gullah" Islands) off the coast of South Carolina. Also in 1867, in an article published in *The Atlantic Monthly*, Col. Thomas Wentworth Higginson (himself a Unitarian minister and abolitionist) wrote about the songs he heard among the black soldiers of his unique, African American Union regiment and among the displaced persons camps (Higginson 1969). Whether the notion that these songs were preserved through relative isolation was an entirely valid belief or not, much of the inspiration for the earliest collections was precisely this belief in the nineteenth century that the "seclusion" of the Sea Islands preserved older versions of African American traditions.

I am therefore intrigued with the possibilities of examining other more "isolated" island cultures with questions about what theological and even Scriptural treasures may be appreciated anew. Macaskill's essay is a trenchant warning against quick dismissal of isolated locations.

Finally, I am among the many (including many of my students) who have benefited greatly from Middleton's book, *The Liberating Image* (2005), so it was a pleasure to read his reflections on Scripture that are built upon a Jamaican background. This is a background that I confess I was not sufficiently conscious of in reading through his book. In this case, however, context is critically important. As a theologian able to reflect on the environmental degradation of his own island of Jamaica, Middleton shares a vision for a "grounded" environmental or "creation" theology. On an island, it is harder to place environmental disaster far from the residents. I doubt many Los Angeles residents are as immediately aware of the dangers of eating some of the fish from the Santa Monica Bay as Middleton is aware of the dangers of fish from Kingston Harbor. Some of us are; most are not. Furthermore, as a Los Angeles resident of twenty-five years, I confess that I have not the faintest idea where the garbage dumps are! I suspect on most islands, however, almost everyone knows where the garbage is! Therefore, islanders remind us all of the importance of being aware of your space—it is not as big as you think! I remember well the earliest shock expressed in media (television ads, newspaper reprints, posters that sold very well) when we saw the photos from the moon—the photos of "island earth" and for the first time having the same sense of immediate awareness of both danger and beauty—and limitations—that Middleton describes in his opening paragraphs. The legacy of the western move in the United States—the vast expanses that one still sees driving through California, Arizona, and New Mexico—continues to lull some of us into a false sense of endless land ready for exploitation ... but California is suffering from unprecedented drought in the opening decades of the twenty-first century, and the Jamaican theologians' warnings are well heeded.

While Middleton's proposed analysis of the biblical basis for a much needed creation theology are, as one would expect, well articulated and argued, what I find the most fascinating is that the *urgency* of his biblical analysis is arguably driven by an island theologians' eye for island responsibilities. It is what makes the argument urgent, but it is also what makes the argument accessible by design. Arguably, Middleton shows us an island theologians' conscience that seeks to make the resulting work popularly accessible—for the pews and the streets and not just for the academies. Middleton's urgency and also his call for accessible argumentation with Scripture in the lingua franca—the creole, if you will—of a people who value biblical "bases" for ethical discussion, is interesting. This insistence that our arguments be accessible brings us right back to where we began with Miller's call for communal processing of biblical thought.

What these essays leave us with at this stage is clear and exciting. First, there is clear significance to the suggestion that Islander Exegesis represents a demonstrably significant perspective. Second, it is equally clear that as islands are numerous, so too are the tasks ahead. Third, the very diversity of different island cultures means that unique treasures of thought and experience are to be shared from different contexts. These essays establish clear lines for future research projects and participation in those projects.

I have a sense of appreciation that I have been invited to listen in on this accomplishment of a clearly mapped course for sailing among islands in reading Scripture. I am furthermore reminded of two collections of essays that marked an "arrival" of sorts—a maturing and confidence in exploring methodological issues related to culturally informed approaches to Scripture, namely, the collection of African American essays in *Stony the Road We Trod* (1991) and the collection of Asian-American essays in *Ways of Being, Ways of Reading* (2006). This volume will take its place next to these critically important watershed collections of essays.

Works Cited

Allen, William Francis, Charles Pickard Ware, and Lucy McKim Garrison, eds. 2011. *Slave Songs of the United States*. Chapel Hill: University of North Carolina Press.

Binney, Judith, 1997. *Redemption Songs: A Life of Nineteenth-Century Maori Leader Te Kooti Arikirangi Te Turuki*. Honolulu: University of Hawaii Press.

Elsmore, Bronwyn. 1985. *Like Them that Dream: The Maori and the Old Testament*. Auckland, NZ: Tauranga Moana.

Felder, Cain Hope. 1991. *Stony the Road We Trod: African American Biblical Interpretation*. Minneapolis: Fortress.

Foskett, Mary F., and Jeffrey Kah-Jin Kuan, eds. 2006. *Ways of Being, Ways of Reading: Asian American Biblical Interpretation*. Saint Louis: Chalice.

Higginson, Thomas, 1969. "Negro Spirituals" Pages 11–22 in *The Social Implications of Early Negro Music in the United States*. Edited by Bernard Katz. New York: Arno.

Horsley, Richard, ed. 1997. *Paul and Empire: Religion and Power in Roman Imperial Society*. London: T&T Clark.

———. 2002. *Jesus and Empire: The Kingdom of God and the New World Disorder*. Minneapolis: Fortress.

Jennings, Steven C. A. 2007. "'Ordinary' Reading in 'Extraordinary' Times: A Jamaican Love Story." Pages 49–62 in *Reading Other-Wise*. Edited by Gerald West. Semeia Studies 62. Atlanta: Society of Biblical Literature.

Middleton, J. Richard, 2005. *The Liberating Image: The Imago Dei in Genesis 1*. Grand Rapids: Brazos.

Writing from Another "Room-in-ating" Place

Randall C. Bailey

It is exciting to be involved on the beachhead of another approach to biblical interpretation. While I have not lived on an island, most of my career as a biblical scholar has been spent on an island inhabited by few people. Often my *talanoa* (storytelling) appearing in edited volumes are washed away into the sea by reviewers who ignore them. Even my *Kāinga*, my kindred in biblical studies, look away as I provide new readings. I am not complaining. Rather, I am posing how I was able to resonate with the contributions of Steed Vernyl Davidson, Nāsili Vaka'uta, and Jione Havea as they contoured what it means to write from an island perspective. What does it mean to write from a place where the empire says nothing can come from that place? What does it mean to write from places surrounded by water, which give its inhabitants different eyes to see and different values from which to explore? What does it mean to read and interpret materials that originate from a totally different geographical space and a totally different set of connections from one's own? Does one do the limbo as a reenactment of the middle passage of training for the discipline? These are constructs raised by the introduction to this volume and the articles contributed by the Hebrew Bible scholars noted above.

In the opening chapter to this volume, Davidson, Margaret Aymer, and Havea contour the project first from the geographical and disruptive nature of islandness. They argue that there is a Western fantasy about islands and that the task ahead is to display how biblical scholarship from the so-called "main land" has ignored the disruptive and alternative approaches to the discipline brought by islanders. The scholars in this volume embrace the geography of islands speaking to topography in terms of the relationship of water to land and sand. They also note how there are islands that imprison people but are not included in the constructs of islander readings. They raise the question of the impact of the scholar's geographical location on one's scholarship. Since islands are insular they raise the possibility of readings from these places also being insular. They explore postcolonial thought from Gayatri Spivak and Kamau Brathwaite and argue that postcolonial theories and marginalization have impacts on island readings of texts, though not all islands were colonized. They conclude that island readings are built upon relationships among people.

Davidson writes his chapter from his Caribbean island experience. Thus, there is an African diasporan contouring to his reading. He uses Stuart Hall to reformat Homi Bhabha's hybridity into a creolization that is a process of creating something new and different from the originating constructs. From his Caribbean place he critiques Kortright Davis's negative view of sandscapes and argues that sand is a transition point brought out of the land by the sea. He, thus, posits a doubleness and a multivalent use of the metaphor of sand. This leads him to claim that island hermeneutics sees the Bible as an import to the island in the same way transposed islanders are and thus to read texts and characters in the text as islanders, whether they are displaced from their homelands or are residing in their own homelands. Utilizing the construct of dislocation for those in the Caribbean leads islanders to seek doubleness in texts. For example, in Gen 38 one focuses on Judah who experiences death and life of his sons. Similarly Tamar is viewed both as being wronged by Judah and also being the mother of two sons. Concentrating on the ebb and flow of tides on the shores of islands leads islanders to look for patterns of motion in the text that either repeat or contradict each other. This moves the reading away from looking for linear time trajectories in the text. Similarly, island life pushes readers not to look for utopian constructs but for possibilities of vulnerability and limitations. Thus, conflicts within the text are expected to be there in line with the realities of life on the Caribbean islands. In line with this, binaries are rejected in favor of multiple possibilities, which help or constrict but which are not ultimate. As opposed to dominance as presented in apocalyptic literature in Daniel, islanders see both the ram and goat experience as positive and negative experiences. Therefore, one should not be looking for good triumphing over evil. Rather one should see them coexisting.

It is interesting that Davidson does not see the sexualizing and exploitation of Tamar as what happens to island/indigenous women by invaders and patriarchal systems. Similarly, he does not connect the levirate practice in Gen 38 with its practices in African and African diasporan contexts. So how does one in Caribbean contexts deal with childless widows? Finally, the relation of Gen 38 to its surrounding context, with Joseph being sold by his brothers in Gen 37 and being bought by Potiphar in Gen 39, makes Gen 38 an island floating around in this part of the book, not really connected to its surroundings but problematizing Judah, who goes to another land, and Tamar, who is used to give a beginning to the Davidic line across the seas of Genesis to 1 Samuel.

In his chapter, Vaka'uta explores relational concepts within Oceania societies as a way to contour how biblical interpretation should be constructed. His concern is mitigating the ways in which Eurocentric hegemonic biblical scholarship has silenced other voices that need to be heard. Thus, he argues for "island marking of texts," which refers to reading in ways which grow out of his culture. Unfortunately, in the chapter Vaka'uta does not apply these concepts to the biblical text itself. He then argues against adopting postcolonial methods, since there are islands which are still colonized. Similarly he rejects ecological approaches, since the task for islanders in Oceania is to survive the ravages of the sea and nature rather than to save the planet.

He then lists and explains values in his society that should be utilized in exploring texts, such as *fale*, which addresses group consciousness. *Kāinga* speaks to that which is applicable to the community. *Tauhi vā* stresses mutual sharing over domination. *Faka'apa'apa* addresses mutual respect, especially in situations of disagreement. *Femolimoli'i* stresses sharing of resources with others, and *fua kavenga* speaks to fulfilling one's obligations to others. He argues that biblical scholarship usually does not employ such values. He concludes that one should read the text with respect and in line with those who are marginalized. He does not, however, address how one engages biblical texts in their construction of narratives and speeches that trample on the Tongan values advocated in this essay. For example, could Abraham be shown to have followed these values? Does Jephthah conform to them? If not, what does one do with such texts? It would have been helpful for him to share islander reactions to such troubling texts.

Havea also takes his lead from value constructs in his island homeland. He argues for *talanoa*, stories and narratives, which can be read together for similarities and differences in plot, especially if Eurocentric scholarship does not connect such *talanoa*. As opposed to narrative criticism, which seeks to hold plots together along narrative lines, *talanoa* allows for connecting stories that reverberate in the eyes of the reader. He thus goes against the grain of reading Ruth and Esther together, as narratives of women living in other lands, and instead reads Ruth and Joseph's mistress[1] in Gen 39, as having interesting connections. Similarly, he does not feel one should rely on narrative nor historical moorings of the *talanoa* within the text. For example, while the narrators of both the Ruth and Joseph texts do not explore the problems of migration in both narratives, as a migrant on an island himself, Havea brings these aspects to play in his exploration of these two narratives, since they are both migrants. He reads both Ruth and Joseph's mistress as positively seeking sexual gratification, and thus he rejects feminist claims of their being abusive and abused. Interestingly, he notes that on the threshing floor in Ruth 3 Boaz assumes the role of Naomi in giving Ruth instructions on what to do both sexually and in terms of addressing the levirate marriage. Havea contrasts Ruth who gleans in fields and Joseph who cleans houses. So there is a playfulness in this approach. He also compares Joseph's mistress with Pharaoh in Gen 12 who "takes Sarai as a wife."

There are two facets of this reading that intrigue me as coming from a different social location than Havea. He does not engage the sexualizing of the Africans, Pharaoh in Gen 12 and the slaveholder in Gen 39, and the exploitation of enslaved bodies by these people. In fact, while he sees the woman in Gen 39 as upper class, he does not see the narrative in line with upper class people sexually exploiting the underclass on the island. Similarly, he does not connect this woman with Hagar in Gen 16 who is another Egyptian woman who is enslaved and sexually exploited. The other striking aspect of his argument is that, while he thoroughly engages the scholarship of European, Israeli, and Euro-American scholars in regard to these units, he does not engage the scholarship of biblical scholars of color who have written on these units of Esther, Ruth, and Joseph.[2]

Having been involved in the development of contemporary Afrocentric biblical interpretation and being asked to enter into dialogue with the writers of this volume sends me back to the pre–*Stony the Road We Trod* (1991)[3] days, when we as black biblical scholars came together to

---

1. He rejects addressing this unnamed character as Mrs. Potiphar. He also relies on Alter's translations of vv. 10b and 14a as suggesting that Joseph was purchased as a slave to sexually satisfy this woman and her charge was that he did not do such as he was supposed to do.

2. For example, while he does reference a discussion he and I had on Gen 39, he does not engage the written works of Donaldson 1993; Bailey 1994, 2009; Gafney 2010; and Yee 2009.

3. The group that generated that volume met for three summers at Collegeville,

hash out what we meant by "black biblical interpretation." As I read these works, I thought back to how we sat at tables and approached the task all coming from different directions. As we shared our own stories of formation and read each other's works, looking for points of connection, we had to explore different usages of method and different interpretations of the text. Like the writers in this part of the current volume, our concerns were how we respond to Eurocentric male interpretations and bring to the table our own people's ways of negotiating the text. So, just as Vaka'uta contours values from his island culture as ways of engaging the tasks of interpretation, I was reminded of the works of Thomas Hoyt who advocated "imagination" and William Myers who struggled with the dissonance of Germanic constructs and black religious interpretations. Just as Havea mixes narratives in different ways not being tied to narrative nor historical contexts, I am reminded of Renita Weems and David Shannon as they deal with asking different questions of biblical texts and argue for new ways of naming dots and new ways of connecting those dots. So, in my experience there is a repertoire for engaging new ways of approaching and contouring biblical studies, which arise from a different "room" than those who have held the keys to boats which bring the text to the island.

While these writers all arise from islands different from each other and approach their understanding of islandness from the different spaces of their places of origin, I was also reminded of my experience with the group which produced the volume *They Were All Together in One Place: Toward Minority Biblical Criticism* (2009). Unlike that group, the writers of this volume did not attempt to learn the "canons within the canon" of other islander scholars and other island cultures. Rather they wrote from the geographical spaces and social cultures of their own social locations. So the secondary task would have to be done by later readers of these works, as Havea invites readers of his article to do. In some ways, the opening chapter to the volume names points of connection and difference in the approaches and contributors to the book on a preliminary level; however, the writers of the pieces themselves do not show how this impacts their own works.

Minnesota, at the Ecumenical Institute at Saint Johns College through a grant from Lily. Some approached the task as black *biblical* scholars. Others saw themselves as black biblical *scholars*. Then there were those who envisioned themselves as *black* biblical scholars. Just as the writers in this volume see themselves functioning from different points of interest and focus, so also did we.

As noted in the introductory essay to the volume, there are connections between island existence and postcolonial discourse, especially for those islands that have historical connections with the various empires. In this way Davidson uses constructs that grow out of the experiences of enslavement of Africans in the Caribbean as a point of departure for his grappling with islandness and how these experiences favor creolization and emancipation over Bhabha's constructs. On the other hand, Havea and Vaka'uta return to Pacific Island culture and language to construct their modes of interpretation. As noted above what is most interesting is that these writers engage more the works of Eurocentric biblical scholars in their constructions than they utilize the works of Asian, black, First Nation, or Latino/a biblical scholars. Part of this could be explained by the guild's having produced and published more scholars from the controlling groups, but there seems not to be a quest to engage these racial ethnic constructs. On the flip side of the water currents going out, they do not show how Eurocentric biblical scholars did the same things in developing their own constructs. Thus, Hermann Gunkel uses Grimm's nursery rimes to develop his form critical constructs. Martin Noth uses the formation of German tribal entities into a nation in his argument for amphictyony. In other words, the moves made by these scholars in forming islander readings based upon geographical and historical realities and values on their islands mimic what scholars from dominating cultures have phenomenologically done in their own scholarship.

Similarly, while there is awareness of engaging feminist concerns, the intersectionality does not move to issues of race, class, and varieties of sexuality. As Audre Lorde (1995) argues, we must engage varied forms of interpretation and various forms of oppression as they intersect, since these points of contact broaden and enrich our interpretations. By the same token, as Mary Ann Tolbert (1990) contours phases of United States feminist interpretation and as Randall Bailey (2000) does the same for United States academic Afrocentric interpretation, one need not begin a new method monofocally. In other words, even in the formation of a new method of interpretation, there are points of contact with other methods beyond Eurocentric methods, which could enrich the discourse. While these writers do engage postcolonial and feminist scholars, the contouring of islander readings could be advanced as they engage broader constructs, which grow out of the struggles of other marginalized groups of biblical scholars who have gone before them.

Finally, while there has been privileging of postcolonial, feminist, and ideological critical interpretive tools within the discipline, there has not been engagement in these articles with newer forms of criticism such as queer theory. For example, Davidson does not engage the eroticism in the Daniel passage. Similarly, while Havea connects Joseph with Ruth, Boaz with Naomi, and the unnamed woman in Gen 39 with Pharaoh, he does not deal with constructs of gender performativity as presented by Judith Butler (2006). Thus, he does not see the feminizing of Joseph, nor the portrayal of Naomi in the pimp role in Ruth 3. Similarly, he does not engage Joseph and Potiphar's relationship as "attending" in line with Abishag and David in 1 Kgs 1:4. In other words there is a queering going on in these texts, which is ignored in Havea's *talanoa*. It is not clear as to whether there was a decision to engage biblical methods which have a noncontroversial edge in their island contexts, such as postcolonialism, but to not engage queer readings that might not be acceptable in those contexts.

Again, I applaud the writers in this volume as they forge ahead in developing and contouring islander hermeneutics. I also thank them for inviting me into the conversation, and I encourage them to be in more dialogue with other biblical scholars from marginalized groups and to engage more newer forms of biblical interpretation for that engagement will not only help them and us but also the profession.

Works Cited

Bailey, Randall C. 1994. "They're Nothing but Incestuous Bastards: The Polemical Use of Sex and Sexuality in Hebrew Canon Narratives." Pages 121–38 in *Social Context and Biblical Interpretation in the United States*. Vol. 1 of *Reading from This Place*. Edited by Fernando Segovia and Mary Ann Tolbert. Minneapolis: Fortress.

———. 2000. "Academic Biblical Interpretation among African Americans in the United States." Pages 696–711 in *African Americans and the Bible: Sacred Texts and Social Textures*. Edited by Vincent Wimbush. New York: Continuum.

———. 2009. "That's Why They Didn't Call the Book Hadassah! The Interse(ct)/(x)ionality of Race/Ethnicity, Gender and Sexuality in the Book of Esther." Pages 227–50 in *They Were All Together in One Place? Toward Minority Biblical Criticism*. Edited by Randall C. Bailey, Tat-

siong Benny Liew, and Fernando F. Segovia. SemeiaSt. Atlanta: Society of Biblical Literature.
Bailey, Randall C., Tat-siong Benny Liew, and Fernando F. Segovia, eds. 2009. *They Were All Together in One Place: Toward Minority Biblical Criticism*. SemeiaSt. Atlanta: Society of Biblical Literature.
Bhabha, Homi K. 1994. *The Location of Culture*. New York: Routledge.
Brathwaite, Kamau. 2004. *Words Need Love Too*. Cambridge: Salt.
Butler, Judith. 2006. *Gender Trouble: Feminism and the Subversion of Identity*. New York: Routledge.
Donaldson, Laura. 1993. "Cyborgs, Cyphers and Sexuality: Re-Theorizing Literary and Biblical Characters." *Semeia* 63:81–96.
Gafney, Wil. 2010. "Ruth." Pages 249–54 in *The Africana Bible: Reading Israel's Scripture from Africa and the African Diaspora*. Edited by Hugh R. Paige Jr. Minneapolis: Fortress.
Hall, Stuart. 1995. "New Cultures for Old." Pages 175–214 in *A Place in the World? Places, Cultures and Globalization*. Edited by Doreen Massey and Pat Jess. Oxford: Oxford University Press.
Hoyt, Thomas, Jr. 1991. "Interpreting Biblical Scholarship for the Black Church Tradition." Pages 17–39 in *Stony the Road We Trod: African American Biblical Interpretation*. Edited by Cain Hope Felder. Minneapolis: Fortress.
Lorde, Audre. 1995. "Age, Race, Class, and Sex: Women Redefining Difference." Pages 532–40 in *Race, Class, and Gender: An Anthology*. Edited by Margaret L. Anderson and Patricia Hill Collins. 2nd Edition. Boston: Wadsworth.
Myers, William. 1991. "The Hermeneutical Dilemma of the African American Biblical Student." Pages 40–56 in *Stony the Road We Trod: African American Biblical Interpretation*. Edited by Cain Hope Felder. Minneapolis: Fortress.
Shannon, David T. 1991. "'An Antibellum Sermon': A Resource for an African American Hermeneutic." Pages 98–123 in *Stony the Road We Trod: African American Biblical Interpretation*. Edited by Cain Hope Felder. Minneapolis: Fortress.
Spivak, Gayatri Chakravorty. 2009. "Nationalism and the Imagination." *Lectora* 15:75–98.
Tolbert, Mary Ann. 1990. "Protestant Feminists and the Bible: On the Horns of a Dilemma." Pages 3–23 in *The Pleasure of Her Text: Feminist Readings of Biblical and Historical Texts*. Edited by Alice Bach. Philadelphia: Trinity Press International.

Yee, Gail A. 2009. "'She Stood in Tears Amid the Alien Corn': Ruth, the Perpetual Foreigner and Model Minority." Pages 119–40 in *They Were All Together in One Place? Toward Minority Biblical Criticism*. Edited by Randall C. Bailey, Tat-siong Benny Liew, and Fernando F. Segovia. SemeiaSt. Atlanta: Society of Biblical Literature.

Weems, Renita J. 1991. "Reading *Her Way* through the Struggle: African American Women and the Bible." Pages 57–77 in *Stony the Road We Trod: African American Biblical Interpretation*. Edited by Cain Hope Felder. Minneapolis: Fortress.

Contributors

Margaret Aymer (maymer@austinseminary.edu) is Associate Professor of New Testament at Austin Presbyterian Theological Seminary. Originally born on the island of Barbados, Aymer has roots in such "roomy nations" as Jamaica and Antigua and Barbuda and was for many years a subject of the imperial archipelago of Great Britain. Her ruminations on migrant writings can also be read in her essay "Rootlessness" in the *Fortress Commentary on the New Testament* (Fortress, 2014) and in *James: Diaspora Rhetorics of a Friend of God*, her contribution to the Phoenix Guides to the New Testament (Sheffield Phoenix, 2015). She has also published *First Pure, Then Peaceable: Frederick Douglass, Darkness, and the Epistle of James* (T&T Clark, 2008), as well as essays in the *Women's Bible Commentary* (Westminster John Knox, 2013) and *Mother Jones, Mother Goose, Mommie Dearest* (Society of Biblical Literature, 2009). In addition to this volume, Aymer is a coeditor of the *Fortress Commentary on the New Testament*.

Randall C. Bailey (rcbitc@gmail.com) is Andrew W. Mellon Distinguished Professor of Hebrew Bible at The Interdenominational Theological Center. He is the author of *David in Love and War: The Pursuit of Power in 2 Samuel 10–12* (JSOT Press, 1990), the editor of *Yet with a Steady Beat: Contemporary U.S. Afrocentric Biblical Interpretation* (Society of Biblical Literature, 2003), and coeditor of *The Recovery of Black Presence: An Interdisciplinary Exploration* (Abingdon, 1995), *Race, Class and the Politics of Bible Translation* (Society of Biblical Literature, 1996), *The Africana Bible: Reading Israel's Scriptures from Africa and the African Diaspora* (Fortress, 2010), and *They Were All Together in One Place? Toward Minority Biblical Criticism* (Society of Biblical Literature, 2009).

Roland Boer (roland.t.boer@gmail.com) is Xin Ao International Professor of Literature at Renmin University of China, Beijing, and Research Professor at the University of Newcastle (Australia). He is the founder of

The Bible and Critical Theory (journal and seminar), serves as one of the managing editors of *Critical Research on Religion*, and continues to be a prolific writer and blogger. Boer recently completed a seminal five-volume series with Brill and Haymarket in which he offers critical commentary on the interactions between Marxism, theology, politics, religions, and living: *Criticism of Heaven* (2007), *Criticism of Religion* (2009), *Criticism of Theology* (2011), *Criticism of Earth* (2012), and *In the Vale of Tears* (2013).

Steed Vernyl Davidson (sdavidson@mccormick.edu) is Associate Professor of Hebrew Bible/Old Testament at McCormick Theological Seminary in Chicago. Prior to this, he taught at Pacific Lutheran Theological Seminary and the Church Divinity School of the Pacific (Berkeley) and served on the Core Doctoral Faculty of the Graduate Theological Union (Berkeley). His current research and teaching interests include Hebrew Bible and ancient Near Eastern empires; postcolonial biblical hermeneutics; Jeremiah studies; early Persian period studies; exile, diaspora and displacement studies and the Bible; and textuality and formation of the Hebrew Bible. He has published *Empire and Exile: Postcolonial readings of the Book of Jeremiah* (T&T Clark, 2011) and several articles and book chapters. He was born on the island of Tobago, part of the twin-island republic of Trinidad and Tobago.

Jione Havea (jhavea@gmail.com) is a native of Tonga who is a principal researcher with the Public and Contextual Theology Research Centre of Charles Sturt University (Australia). Jione is a Methodist minister who fishes for island and contexted readings with queer goggles and native hands and enjoys grog among relaxed and rewinding company. He is the author of *Elusions of Control: Biblical Law on the Words of Women* (Society of Biblical Literature, 2003) and coeditor of *Out of Place: Doing Theology on the Crosscultural Brink* (Equinox, 2011).

Hisako Kinukawa (hisakokinukawa@gmail.com) teaches at theological seminaries and works with two groups that focus on women and theology: the *Asian Women's Resource Center for Culture and Theology*, currently based in Indonesia, and the *Center for Feminist Theology and Ministry* in Japan. This work has helped shape her strong commitment to nurturing her theology as an Asian woman. The divisiveness and conflict arising around the small islands surrounding Japan make her wonder how the "property" concept of the earth evolved. Perhaps clues lie in the book of

Genesis in the Hebrew scriptures. She is the author of *Women and Jesus in Mark: A Japanese Feminist Perspective* (Orbis, 1994) and editor of *Migration and Diaspora: Exegetical Voices from Northeast Asian Women* (Society of Biblical Literature, 2014).

Mosese Ma'ilo (mosemailo@yahoo.com) resides and lectures in biblical studies (New Testament and postcolonial biblical hermeneutics) at Piula Theological College, Samoa. Ma'ilo is the current president of Oceania Biblical Studies Association and holds a PhD in Postcolonial Bible Translation from the University of Birmingham, United Kingdom. He is also an ordained minister of the Methodist Church in Samoa.

Grant Macaskill (gm37@st-andrews.ac.uk) is Senior Lecturer in New Testament at the School of Divinity, University of St Andrews (Scotland). He is the author of several monographs, including *Union with Christ in the New Testament* (Oxford University Press, 2013) and *The Slavonic Texts of 2 Enoch* (Brill, 2013). His teaching areas include Jesus and the Gospels, Johannine literature, the book of Revelation, New Testament Christology and pneumatology, biblical exegesis, the Old Testament pseudepigrapha, Enochic Judaism, and theological interpretation of scripture.

Andrew Mein (arm32@cam.ac.uk) is Senior Research Fellow in Biblical Studies at Westcott House, Cambridge, and has taught Hebrew Bible/Old Testament in the Cambridge Theological Federation and the University of Cambridge since 1997. His main research interests focus on the book of Ezekiel, the ethics of the Hebrew Bible, and the reception and influence of the Bible in culture and society. He is the author of *Ezekiel and the Ethics of Exile* (Oxford University Press, 2001).

J. Richard Middleton (Middleton_Richard@roberts.edu) is Professor of Biblical Worldview and Exegesis at Northeastern Seminary, in Rochester, New York. A native of Jamaica, Middleton earned his first theological degree from Jamaica Theological Seminary and serves as visiting professor of Old Testament at the Caribbean Graduate School of Theology in Kingston. He is the author of *A New Heaven and a New Earth: Reclaiming Biblical Eschatology* (Baker Academic, 2014) and *The Liberating Image: The Imago Dei in Genesis 1* (Brazos, 2005). He has coedited *A Kairos Moment for Caribbean Theology: Ecumenical Voices in Dialogue* (Pickwick, 2013) and is coauthor of *The Transforming Vision: Shaping a Christian World*

View (InterVarsity Press, 1984) and *Truth Is Stranger Than It Used to Be: Biblical Faith in a Postmodern Age* (InterVarsity Press, 1995).

Althea Spencer Miller (aspencer@drew.edu) joined the Drew Theological School faculty in 2008 and is Assistant Professor of New Testament Studies there. Postcolonial hermeneutics, flavored by anticolonial interests in orality studies and Africana studies, frame her research interests. Feminist, queer, and womanist theories supplement these. Spencer-Miller cochairs the "Islands, Islanders, and Bible" Society of Biblical Literature program unit, collaborates with the African Association for the Study of Religion, and is on the editorial board of *Sapienta Logos Journal* (Nigeria). Her publications include a coedited volume on *Feminist New Testament Studies: Global and Future Perspectives* (Palgrave Macmillan, 2005); "Women and Christianity in the Caribbean: Living Past the Colonial Legacy," in *Women and Christianity* (Praeger, ABC-CLIO, 2010); and "Rethinking Orality for Biblical Studies," in *Postcolonialism and the Hebrew Bible* (Society of Biblical Literature, 2013).

Aliou C. Niang (aniang@uts.columbia.edu) is Assistant Professor of New Testament at Union Theological Seminary in New York. He is author of *Faith and Freedom in Galatia and Senegal: The Apostle Paul, Colonists and Sending Gods* (Brill, 2009) and coeditor of *Text, Image, and Christians in the Graeco-Roman World: A Festschrift in Honor of David Lee Balch* (Pickwick, 2012). He is currently working on a book project on the Gospel of Mark.

Daniel Smith-Christopher (Daniel.Smith-Christopher@lmu.edu) is Professor of Old Testament Studies and Chair of the Graduate Program in Theology at Loyola Marymount University in Los Angeles, where he has taught for twenty-five years. Smith-Christopher has a particular interest in crosscultural interpretation of Scripture, especially in relation to minority and indigenous traditions of Christianity. Smith-Christopher works with Rev. Donald Tamihere on issues related to biblical interpretation and Maori cultural perspectives in New Zealand. Smith-Christopher is completing a commentary on Micah and is working on a third edition and revision of the late Fr. Anthony Ceresko's *Introduction to the Old Testament: A Liberationist Perspective* (Orbis).

Nāsili Vakaʻuta (nvakauta@gmail.com) is the Principal of Trinity Methodist Theological College and is one of the founders of Oceania Biblical Studies Association. He also taught at Sia'atoutai Theological College, Tonga, and is committed to the cause of natives, especially the *tu'a* (commoners), as well as *moana* (sea) and *fonua* (land), as he reads and interprets biblical texts. He is author of *Reading Ezra 9–10 Tu'a-Wise: Rethinking Biblical Interpretation in Oceania* (Society of Biblical Literature, 2011).

Elaine M. Wainwright (em.wainwright@auckland.ac.nz) has retired after serving as Head of the School of Theology at the University of Auckland, New Zealand. She is a New Testament scholar committed to exploring how ancient texts function within contexts of interpretation within the contemporary world. She is coeditor of *The Bible in/and Popular Culture: A Creative Encounter* (Society of Biblical Literature, 2010) and author of *Shall We Look for Another: A Feminist Re-reading of the Matthean Jesus* (Orbis, 1998), *Women Healing/Healing Women: The Genderisation of Healing in Early Christianity* (Equinox, 2006), and *The Gospel according to Matthew: The Basileia of the Heavens Is Near at Hand* (Sheffield Phoenix, 2014). Her current focus is ecological hermeneutics, and she is writing the Earth Bible Commentary on Matthew's Gospel.

Index of Ancient Texts

Hebrew Bible/Old Testament

Genesis
Reference	Page
1:1	121
1:26–28	122, 125
1:31	122
2–3	111
2:15	122
3	124–25
4–11	124
4:8	124
4:17	124
4:20–22	124
4:23–24	124
6:11	124
11:31–12:4a	152
12	220
12:3	33
12:10	152
12:10–20	157
16	220
20:1–18	158
22:18	33
26:1–16	158
37	219
38	44, 48, 153, 169, 218–19
38:1	48
38:3	48
38:4	48
38:5	48
38:12	48
38:24	48
38:26	44
38:27	48–49
39	154–55, 219–20, 223
39:5b	155
39:6b	156
39:7	157
39:7b	149
39:10	157
39:17–20	157
42:1–5	152

Joshua
Reference	Page
2:15	15

Ruth
Reference	Page
1:6	151
1:6–7	151
1:8	151
2:2–3	152
2:6	152
2:8	152
2:9	152
2:10	152
3	223
3:1	153
3:3–4	153
3:8	153
3:10–11	153
3:13	154
3:14	154
3:15	154
3:18	154
4:6	158

1 Samuel
Reference	Page
8	14

INDEX OF ANCIENT TEXTS

Ezra
9:2 — 211

Nehemiah
13:24 — 211

Esther
4:14 — 154

Psalms
6:5 — 122
8:4–8 — 125
8:5–8 — 122
18:23 — 104
22 — 104
29 — 104
29:5 — 104
30:9 — 122
36:6 — 108
36:7 — 108
40 — 109
40:2 — 109
47 — 194
63 — 104
68:4 — 129
72 — 190, 193
72:10–11 — 189, 194
77 — 104, 108
77:19 — 191
84 — 104
88:3–5 — 122
88:10–12 — 122
96 — 123
103 — 104
113–118 — 107
115:17 — 122
120–134 — 107
121:1 — 104
122 — 107
126 — 104
148 — 123

Proverbs
3:18 — 111

Isaiah
23 — 139
25:6–8 — 125
38:9–12 — 122
38:18 — 122
65:17 — 121
65:17–25 — 121
66:22 — 121

Jeremiah
47:4 — 139
51:45 — 33

Ezekiel
27 — 139
27–28 — 170
27:3 — 139
27:17 — 140
28 — 139
37 — 125

Daniel
8 — 51, 53, 169–70
12:2–3 — 125

Joel
3:4–8 — 139

Zechariah
9:2 — 139

New Testament

Matthew
1:1–14 — 72, 180
1:18 — 70
1:18–25 — 69, 71, 73, 180, 204
1:18–26 — 66, 204
1:20 — 70
1:23 — 70–71, 180
5:5 — 125
7:24–27 — 168
12:36–50 — 72
19:23 — 121
19:27–30 — 126

INDEX OF ANCIENT TEXTS

Mark
- 6:3 — 72
- 7:24 — 136
- 7:24–30 — 136, 204
- 7:27 — 137
- 7:28 — 137
- 10:42–45 — 125

Luke
- 1:24 — 70
- 4:18 — 125
- 21:47 — 11

Acts
- 3:21 — 121
- 17 — 211
- 20:27 — 130
- 21:1–3 — 138
- 21:3–6 — 137

Romans
- 8:18–25 — 119
- 8:19–22 — 126
- 8:19–23 — 179
- 12:1–2 — 123

1 Corinthians
- 15:20 — 126
- 15:24–28 — 126
- 15:26 — 125
- 15:49 — 126
- 15:54 — 125

Galatians
- 5:6 — 33

Ephesians
- 1:20 — 121
- 4:13 — 126
- 4:24 — 126

Philippians
- 2:5–11 — 126
- 3:20 — 179

Colossians
- 1:20 — 121
- 3:9–10 — 126

1 Thessalonians
- 4:13–18 — 179

James
- 3:1 — 130

2 Peter
- 3:13 — 121

1 John
- 3:2 — 126

Revelation
- 2:13 — 32
- 2:14 — 33
- 2:20 — 33
- 2:24 — 32
- 5:9–10 — 126
- 11:5 — 126
- 13 — 32
- 13:11–15 — 32
- 18:4–5 — 33
- 21:1 — 47, 121, 179
- 21:4 — 125
- 22:3–5 — 126

QURAN

Yusaf
- 12:30–34 — 156

Index of Modern Authors

Addison, Catherine 4, 20
Allen, William Francis 213, 215
Alter, Robert 156–57, 159
Anzaldúa, Gloria E. 16, 17, 20
Ashcroft, Bill 40, 54
Bailey, Randall C. 2, 18, 20, 156–57, 220, 222–24
Baldacchino, Godfrey 4, 6–7, 20, 26, 36
Barnett, Richard D. 139–42, 145
Barry, Peter 93
Barth, Karl 117, 132
Bassnett, Susan 68, 76
Baugh, Edward 46, 54
Bayoumi, Moustafa 93
Beal, Timothy K. 150, 161
Ben Beya, Abdennebi 90, 93
Benítez-Rojo, Antonio 41, 43, 45, 47–48, 51–52, 54
Bennett-Coverly, Louise 25, 31, 36, 43–44, 178–79
Bernabé, Jean 92, 93
Berry, John 28–29, 36
Bertrand-Bocandé, Emmanuel 177–78
Bertrand-Bocandé, Jean 177–78
Bhabha, Homi 40, 44, 54, 67, 74, 76, 90–91, 93, 169, 180, 182, 218, 222, 224
Binney, Judith 210, 215
Boff, Leonardo 120, 125, 132
Boer, Roland 2, 172, 174, 201
Boeree, C. George 141, 145
Boomer, Arie 28, 36
Brah, Avtar 27–28, 31, 36
Brathwaite, Kamau 7–8, 20, 39, 41–42, 54, 218, 224
Brenner, Athalya 151, 159
Bright, Alistair 28, 36
Brown, Stewart 8, 20
Brueggemann, Walter 118, 132
Brunner, Emil 117, 132
Buchanan, Joni 101, 113
Bush, Frederic W. 150, 159
Butler, Judith 223–24
Césaire, Aimé 77, 82, 85, 93, 182
Chamoiseau, Patrick 92, 93
Champion, Timothy 187, 195
Chomsky, Noam 15, 20
Chow, Rey 40, 54
Chrisman, Laura 91, 94
Cone, James H. 119–20, 132
Confiant, Raphaël 92, 93
Conkling, Philip 4, 8, 20
Cosden, Darrell 123, 132
Crouch, Andy 123, 132
Dash, Michael 80–82, 93
Davis, Kortright 38–39, 45, 49–50, 54, 218
Debien, Gabriel 177, 183
DeLoughrey, Elizabeth M. 3, 5, 7, 20
Desai, Gaurav 93
Donaldson, Laura 220, 224
Dubois, W. E. B. 80, 93
Edmondson, Belinda 93
Elsmore, Bronwyn 210, 215
Eskenazi, Tamara Cohn 150–51, 159
Fanon, Franz 77, 82
Felder, Cain Hope 215
Fewell, Danna N. 1, 20, 149, 151, 153, 159–60
Figiel, Sia 147, 160
Fretheim, Terence E. 119, 124, 132

INDEX OF MODERN AUTHORS

Frymer-Kensky, Tikva 150–51, 159
Foskett, Mary F. 215
Gafney, Wil 220, 224
Garuba, Harry 5, 20
Gikandi, Simon 40–41, 43, 54
Gillis, John R. 26, 36
Gillis, John T. 3–5, 20
Glissant, Édouard 77–78, 80–83, 85–93, 181–83
Goldman, Dara E. 38, 49, 52, 54
Gorman, Michael J. 118–19, 124, 133
Greenstein, Edward L. 150, 160
Griaule, Marcel 90, 93
Gribben, Crawford 192, 195
Griffin, Susan 117, 132
Griffiths, Gareth 40, 54
Gunn, David M. 149, 151, 153, 160
Hager, Nicky 148, 160
Hall, Stuart 40, 54, 218, 224
Hardt, Michael 167, 174
Harris, Wilson 40–43, 54
Hau'ofa, Epeli 57, 64
Havea, Jione 1–2, 7–8, 15, 20, 147–49, 160, 177, 186–87, 202, 207–8, 210, 217, 219–23
Hawn, C. Michael 194–95
Hay, Pete 11–12, 20, 27, 31, 35–36
Heggulund, Jon 5, 20
Hiebert, Theodore 122, 132
Higginson, Thomas 213, 215
Holladay, Carl R. 179, 183
Honig, Bonnie 151, 160
Hooks, Bell 6, 20
Horsely, Richard 212, 216
Hoyt, Thomas Jr. 221, 224
Hubbard, Anthony 148, 160
Hull, John 194, 196
Iyasere, Solomon O. 94
James, Leslie 85–87, 91, 94, 182
Jeansonne, Sharon Pace 155, 157, 161
Jennings, Steven C. A. 210, 216
Kaltner, John 156, 158, 161
Kramer, Augustin 71, 76
Kuan, Jeffrey Kah-Jin 215
LaCocque, André 150, 153, 161
Lamming, George 37, 54
Liew, Tat-Siong Benny 18, 20, 58, 224
Linafelt, Tod 150, 161
Lorde, Audre 222, 224
Lowenthal, David 3, 4, 8, 21
Macdonald, Calum 108–9, 111–12, 113
Macdonald, Rory 108–9, 111–12, 113
MacInnes, John 111–112, 113
MacLeod, John 99, 101, 103, 107–8, 113, 190, 193, 196
Mack, Deborah L. 178, 183
Mackinder, Halford 5
MacNeal, Dean 179, 183
Malinowski, Bronislaw 171–72, 174
Marley, Robert Nesta 25, 29, 36, 129–30, 178–79
McFarlane, Adrian 87, 94
McGee, J. Vernon 150, 161
McKim Garrison, Lucy 213, 215
McLeod, John 68, 76
McNeil, Kenneth 193, 196
Meek, Donald 192, 196
Meeks, M. Douglas 78, 83–84, 86
Mehrez, Samia 66–67, 76, 180
Mein, Andrew 2, 166, 190, 196
Middleton, J. Richard 2, 12, 115, 118–19, 123–124, 129, 133, 165, 172–74, 214, 216
Míguez, Néstor 94
Millar, Fergus 188, 196
Miller, Terry 102–3, 113
Milner, G. B. 70–72, 76
Miscall, Peter D. 186, 196
Moran, J. F. 189–90, 196
Murison, David 192, 196
Myers, William 221, 224
Naipaul, V. S. 37, 54, 80
Nardal, Jane 81
Nardal, Paulette 81, 94
Negri, Antonio 167–68, 174
Niang, Aliou Cissé 2, 178, 183
Nitschke, Jessica 188, 196
Nesbitt, Nick 82–83, 85, 94
Neville, David J. 15, 20
O'Day, Gail R. 139, 145

INDEX OF MODERN AUTHORS

Olwig, Karen Fog 4, 21
Parman, Susan 101–2, 106, 113
Paxman, Jeremy 185, 193, 196
Paz, Octavio 68, 76
Polanyi, Karl 171, 174
Pratt, George 70–72, 76
Pico della Mirandola, Giovanni 117, 133
Rad, Gerhard von 118, 133
Reed, Jonathan L. 136, 145
Revere, Robert B. 170, 174
Roper, Garnett 115, 133
Rose, Gillian 4, 21
Rubin, Andrew 93
Said, Edward 3, 15, 21, 90–91, 93–94, 182
Saint Martin, Yves 177, 183
Sanneh, Lamin 181, 183
Sarna, Nahum M. 157, 161
Scott, Jonathan 187, 197
Scullion, John J. 157, 161
Segovia, Fernando F. 18, 20, 58, 77, 172, 174, 224
Selbo, Jule 171, 175
Sembène, Ousmane 178, 181–183
Senghor, Léopold Sédar 77, 83, 182, 184
Shannon, David T. 221, 224
Smith, Ashley 119, 131, 133
Smith, Wilfred Cantwell 10, 21
Smith-Christopher, Daniel 2, 16, 21
Snyder, James L. 128
Soja, Edward W. 4, 21
Sommer, Doris 94
Spivak, Gayatri Chakravorty 13–15, 18–19, 21, 90–91, 148, 161, 182, 218, 224
Spong, John Shelby 72, 76
Stephens, Gregory 179, 184
Sugirtharajah, R. S. 1, 21, 58, 74–75, 94
Taylor, Patrick 50–51, 53, 54
Theime, John 46, 54
Theissen, Gerd 136–137, 139–41, 145, 188, 197
Tiffin, Helen 40, 54
Tolbert, Mary Ann 222, 224
Tuan, Yi-Fu 6, 8, 21
Vaka'uta, Nāsili 2, 12–13, 16, 147, 161, 165, 170–72, 217, 219, 221, 222
Van Aarde, Andries 71–72, 76
Wainwright, Elaine M. 2, 15
Walcott, Derek 38, 40–41, 45–47, 55, 80, 94
Wallace-Hadrill, Andrew 182, 184
Wallerstein, Immanuel 173, 175
Walsh, Brian J. 123, 133
Ware, Charles Pickard 213, 215
Warner-Lewis, Maureen 88, 94
Weems, Renita J. 221, 225
West, Cornel 59, 64
Westermann, Claus 118, 133
Williams, Patrick 91, 94
Wimbush, Vincent 28, 36
Wolters, A. M. 123, 133
Wright, G. Ernst 118, 134
Wright, N. T. 122, 134
Wright, Stephen 6
Yee, Gail A. 220, 225
Zahan, Dominique 90, 95

www.ingramcontent.com/pod-product-compliance
Lightning Source LLC
Chambersburg PA
CBHW031709230426
43668CB00006B/159